Lecture Notes in Computer Science 2217

Edited by G. Goos, J. Hartmanis, and J. van Leeuwen

Springer

Berlin
Heidelberg
New York
Barcelona
Hong Kong
London
Milan
Paris
Tokyo

Takashi Gomi (Ed.)

Evolutionary Robotics

From Intelligent Robotics to Artificial Life

International Symposium, ER 2001
Tokyo, Japan, October 18-19, 2001
Proceedings

Springer

Series Editors

Gerhard Goos, Karlsruhe University, Germany
Juris Hartmanis, Cornell University, NY, USA
Jan van Leeuwen, Utrecht University, The Netherlands

Volume Editor

Takashi Gomi
Applied AI Systems, Inc.
3232 Carp Road, Carp, Ontario, Canada K0A 1L0
E-mail: gomi@AAI.ca

Cataloging-in-Publication Data applied for

Die Deutsche Bibliothek - CIP-Einheitsaufnahme

Evolutionary robotics : international symposium ; proceedings / ER 2001,
Tokyo, Japan, October 18 - 19, 2001. Takashi Gomi (ed.). - Berlin ;
Heidelberg ; New York ; Barcelona ; Hong Kong ; London ; Milan ; Paris ;
Tokyo : Springer, 2001
    (Lecture notes in computer science ; Vol. 2217)
    ISBN 3-540-42737-6

CR Subject Classification (1998): F.1.1, I.2.9, I.2

ISSN 0302-9743
ISBN 3-540-42737-6 Springer-Verlag Berlin Heidelberg New York

Springer-Verlag Berlin Heidelberg New York
a member of BertelsmannSpringer Science+Business Media GmbH

http://www.springer.de

© Springer-Verlag Berlin Heidelberg 2001

Typesetting: Camera-ready by author, data conversion by PTP-Berlin, Stefan Sossna
Printed on acid-free paper   SPIN 10840884   06/3142   5 4 3 2 1 0

Preface

This volume constitutes the collection of papers presented and debated at the 8th International Symposium on Evolutionary Robotics (ER 2001), subtitled From Intelligent Robotics to Artificial Life, held in Tokyo on the 18th and 19th of October 2001. A paper by Dr. Hiroaki Kitano of Sony Computer Science Laboratory, who also presented at the symposium, unfortunately could not be included because of the severe time constraints under which the invited speakers had to prepare their manuscripts.

Eight years have passed since we first organized our Evolutionary Robotics symposium (ER'93) in Tokyo in April of 1993. During those eight years, we have run a total of eight symposia with the same title and objectives. That itself is rather surprising. Many sophisticated and complex robots have been developed during this period and launched into society. Yet the subjects we decided to study back in 1993 have remained important throughout the research and development community, if not become more important.

We have noticed, through the life of this series of symposia, the existence of some fundamental research themes that seem to demand attention beyond the immediate scientific and technological concerns associated with intelligent robot research. We call such issues philosophical, as philosophy by definition deals with fundamental themes that underlie what can be readily observed using existing science and technology. History shows us that philosophy itself has its own history. We also know that something mankind has come up with through his conscious effort remains effective only for a certain period of time, a few centuries at most (with the notable exception of the Flat Earth Society). It could be that we need to examine the philosophy itself on which today's science and technology are constructed and maintained. If so, unless we focus our attention accordingly and deepen our understanding of key subjects, we will never enjoy the company of the intelligent beings we anticipate producing. Thus it has become a tradition of the symposium series to invite a philosopher or two, as well as those who are philosophically oriented in their daily practice. We have already seen some remarkable results in which revised philosophy has clearly, unquestionably, and considerably improved robots' performance. Incidentally, this year we had two winners of the respected "Computers and Thought" award as invited speakers. However, the philosophical inclination at ER 2001, as well as at our previous symposia, did not stop there. All invited speakers have, through their exceptional careers, deepened their thought and inquiry to the point of being widely praised and accepted as hallmarks in their own sub-disciplines. They have each raised their level of awareness to try to answer some fundamental questions in the field such as "What is computation?", "Where are we heading with our search for intelligence?", "What is intelligence?", and "What is life?"

At the Advanced Study Institute workshop on intelligent robotics held at historic Monte Verita in Switzerland in the fall of 1995, a participant briefly talked about Gray Walter's tortoise robot built around 1950. Owen Holland had just unearthed Walter's fascinating robot built with two vacuum tubes and a few primitive sensors. The concluding session of this two-week intensive workshop spent a considerable amount of time on the impact of the excavation and the impact of Gray Walter's work which had already incorporated *behavior-based* principles, the very subject the workshop was supposed to examine in depth. Participants then asked, "Have we made any progress since Gray Walter?". Today, several years later, and exactly half a century after the tortoise robot, the question is still valid. In his paper at ER 2001, Holland discussed and analyzed the classical work in detail, then turned to the question of *consciousness*. Maybe *consciousness* is totally out of the question at this point, or not so far away as we always think. Who knows?

Dario Floreano has been one of the most active among the researchers who pursue Evolutionary Robotics through experimentation. He has left a trail of incredibly novel and successful experiments since his very first, and the world's first, experiment on embodied and *in situ* evolution of a physical robot in the early 1990s. Recently, he has been focusing on two topics. One is the use of spiking neurons as an element of a robot's autonomy generation mechanism. The other is the use of vision inputs as the principal source of sensory signals to robots. Both topics are very welcome selections as we know that spiking aspects of natural neural networks and the dynamism such networks afford play a significant role in generating the intelligence animals require. In artificial neural networks, such a transient aspect of the network's operation is often ignored and networks with a quasi-steady state processing model are routinely used. Floreano's early experiments also assumed such networks. His initial success in spiking neuron circuits reported here is already very encouraging.

It is similarly exciting that Floreano is concentrating on evolving a robot's ability to handle visual inputs as the dominant source of signals from its operational environment. We already know that vision plays a major role in the sensory processing of a vast number of animals. Technology, as well as methodology to implement scientific findings, had not been quite at par in vision processing until recently. He was successful in obtaining some initial results in this area, as reported here. It is reassuring to see these results, particularly knowing that Floreano always places emphasis on less or no interference from human experimenters while conducting his increasingly sophisticated and always lengthy experiments.

Jordan Pollack attacks the issues of co-evolution in order to study minute yet essential interactions between a robot and its environment. To study such phenomena successfully in artificial evolution, both body and its behavior, or "body" and "brain", as in natural systems, need to be co-evolved. According to Pollack, the interaction between such a co-evolved robot and its environment would result in a unique form of self-organization. This theme was originally very effectively demonstrated by Karl Sims in his pioneering co-evolution study done in a virtual world (shown at ER'94). Pollack pushed the concept further to literally create morphologically evolved physical robots coming out of a co-evolution by a clever use of 3D printing technology. Though the process was very time consuming, he was successful in the automatic manufacturing in plastic of Sims-like creatures and running them after evolving them in virtuality. The success of his experiment drew a lot of attention and was reported in *Nature* in the summer of 2000. Here at ER 2001, Pollack discussed his innovation in the backdrop of the recent world-wide push towards creating increasingly more intelligent robots, and critically reviewed the trend towards their acceptability in society, including their economic viability.

When Rodney Brooks participated in the first ER symposium (ER'93), he was still coping with both positive and negative reactions to his Subsumption Architecture of 1986 coming from the real world as well as the research communities. Also he had just started his humanoid project at MIT's Artificial Intelligence Laboratory. His new intelligent robotics company (now called iRobot) had just one year under its belt with only 2.2 employees, 0.2 being Brooks. He was still a few years away from his directorship of the AI Laboratory. Since then, a lot of things have happened. Under his directorship, the 230-researcher AI Laboratory has regained vitality, and his company has grown rapidly to become the largest company in the world solely dedicated to intelligent robotics, with 125-plus employees. Many projects involving robots are conducted there for a wide range of real-world applications. The survival and growth of his business alone is a testimony, at least in part, to the effectiveness of his theories, most vividly represented by Subsumption Architecture. After chasing *intelligence* for several years, Brooks this time came up with a more fundamental topic, *life*. In a way, *intelligence* itself is an abstraction that our New AI tends to frown upon. The *frame of reference* prejudice could well apply in our effort to seek *intelligence* even though we are very aware of the *problem* and the pitfalls it sets up. Maybe we need to go deeper and investigate what distinguishes *life* from *non-life*, as Brooks pointed out. At ER 2001, Brooks discussed the process of finding the "juice" that makes the difference between the two by examining a few primitive biological systems. *Intelligence* surely must be a side effect of a living organism. By focusing our attention on issues beyond *intelligence*, we may eventually see a light at the end of the *intelligence* tunnel.

Inman Harvey, a Sussex computer philosopher, is also a theoretical big wig in the field of Evolutionary Robotics. Jointly with Phil Husbands and David Cliff, he proposed a new framework to explain evolution, artificial or otherwise, called SAGA (Species Adaptation GAs) in the early 1990s. It was presented at various times at our symposia. Then in 1998, Harvey proposed the concept of Neutral Networks to elaborate events and processes that occur in the course of evolution and most visibly in a fitness landscape. At ER 2001 Harvey attempted to explain Neutral Networks by linking the subject to other highly acclaimed research at Sussex by Adrian Thompson on evolvable hardware. SAGA emphasizes the importance of viewing evolution free of the biases we tend to attach to it, so that we can learn more about evolution. The group has been highly critical of much research on computational evolution where researchers casually and almost routinely adopt a notion of "progress", or "optimization" into their evolution. As such, the concept of SAGA renders itself both controversial and philosophical.

With his extensive background in biology and related fields, his current position as head of a laboratory closely related to Birobotics, and his enviable accumulation of experience in transferring his academic achievements to the real world in a large number of projects already conducted with industry, Robert Full is in a unique and excellent position to overview what biologically inspired artificial systems mean and what their potentials as well as drawbacks are. His paper at ER 2001 documented some of his extensive experience and thoughts in very readable form, summarizing points quite effectively for the far less experienced.

That said, his view on natural systems is not as idealizing or romantic as some might wish. He points out a number of shortcomings in design that biological systems must endure, as well as those in the way they are implemented in their legendary and often mythical functionalities. While the engineering approach for example, here juxtaposed against the biological approach, requires a collapse of dimensions and gives up the incredibly rich, complex, and elegant motions of animals, he stressed that biological evolution has created creatures based on a "just good enough" principle. "Organisms are not optimally designed and natural selection is not engineering," he pointed out. Then he talked about some five hundred million species that became extinct because of basic compromises in design and implementation, leaving only a few million, as natural evolution works more "as a tinkerer than engineer." I could not help but feel some philosophical issues cutting in. Reliance on evolutionary process will not necessarily result in design that is better than a human engineer can do, he also stressed. However, it is now widely accepted that a good part of those five hundred million species became extinct since we began mastering considerably more efficient engineering methodologies, to the point of leaving only a few million to go. What happens next? His view, or my interpretation of his view, opposes that of others in the ER community in several ways, making his study a thought-provoking one.

Regardless, his research is unquestionably interesting and entertaining, particularly when it is presented as crisply as at ER 2001. He even began describing how biologically inspired robots should be constructed.

After making the Khepera robot available to the world's intelligent robot research community, and then launching K-team as a mechanism to handle the distribution, Francesco Mondada took a turn which puzzled many around him. He could have easily achieved a so-called "commercial success" with his talent and prowess. Instead, he directed and consumed a good part of his resources to satisfy researchers in universities and research institutions. He was extremely stubborn about this, keeping product quality high to the point of often sacrificing monetary accountability in the conventional business sense. Many failed to understand his intentions, but I was very fortunate to be one of the few who did, at least to a certain degree. In effect, he began an entirely new trend in robotics and robotic business: robots for the betterment of society in the true and unselfish sense of the words, and business for the same purpose. Despite the objections and constraints the reality around him imposed, he immediately demonstrated his excellence in such an unconventional business framework. He ventured into a few education+entertainment (so called "edutainment") applications. I am reluctant to classify his robots in that category, however, because there is a fundamental difference in the way Mondada plunged into his development and the way he made available the result of his innovation and efforts. Other edutainment and similar robots for the real world are developed and delivered for the direct material gain of the innovator, a common practice in society which culminated in the 20th century. Perhaps my desire to see an idealistic future has clouded my view, but I somehow feel that he was aiming at something a bit more exciting than that, something that is more fit for the 21st century. I sincerely hope you see it that way too.

Acknowledgements. I am greatly indebted to Inman Harvey and Dario Floreano for the successful organization of ER 2001, in particularly, in the selection of speakers.

We, at Applied AI Systems Inc., are very grateful to a number of people and organizations that have supported and maintained the ER symposia so far. The Canadian Embassy in Tokyo has given us generous access to their staff and facility, including an attractive theater and lobby for the gatherings, their complete audio visual system, and simultaneous translation booths. Without such support, finding a venue for this type of gathering in notoriously expensive Tokyo would have been overly painful for this non-profit undertaking.

We are also thankful to the industrial sponsors that gave us financial support to cover the surprisingly vast logistics, most of which are invisible on the surface.

We are also very grateful to all invited speakers who so generously gave their time to prepare materials so that attendees and other readers could be introduced to this exciting new field in advanced robotics. More importantly, we thank them for gladly accepting our invitation and coming all the way to attend the meetings in Japan. Their attendance gave a unique opportunity to many young researchers, in particular for face to face discussion and bouncing of ideas with the world's top researchers in the field of advanced robotics. I am very sure that these were useful in passing on knowledge, experience, and passion to younger generations, as well as to the more seasoned researchers.

We are also thankful to the participants who took the trouble to find sources of financial support or funding and devoted a considerable amount of time to come to Tokyo and spend two days there, either domestically from Japan or from overseas. To some this must have been a major financial burden. We appreciate the courage and effort of those attendees.

October 2001   Takashi Gomi

Table of Contents

From the Imitation of Life to Machine Consciousness

Owen Holland

Intelligent Autonomous Systems Laboratory
Faculty of Engineering
University of the West of England
Bristol BS16 1QY
and
Microsystems Laboratory
California Institute of Technology
Pasadena, CA91125
`owen@micro.caltech.edu`

Abstract. There are many motivations for building robots which imitate life or living things. This paper examines some aspects of the use of robots in the imitation of life for scientific purposes, and divides the field into two parts: the imitation of behaviour, and the imitation of consciousness. A re-examination of the early work of Grey Walter in mimicking behaviour is used as the inspiration for a possible approach to building conscious robots.

1   Introduction

Over the years, one of the strongest motivations within the science of robotics has been to make machines that imitate living things, and in particular machines that imitate animals and humans. This idea is so familiar to us that often we do not analyse it. However, in science, as in other enterprises, we can learn a lot by questioning precisely those ideas that are taken for granted. Let us look at some of the possible reasons for adopting this approach.

1.1   Nature as a Source of Robot Designs

One of the most natural reasons for adopting a strategy of copying living beings is that they often represent successful engineering designs for coping with certain problems. For example, if a robot is required to have good mobility on flat surfaces, combined with the ability to move through narrow gaps, then it is clear that the study of how snakes move might be very useful. (Some idea of the level of world-wide interest in snake robots can be obtained from http://ais.gmd.de/~worst/snake-collection.html, maintained by Rainer Worst.) Interestingly, although we might at first suppose that all snakes move in basically the same way, there turn out to be several different types of snake movement, each with its own characteristics. Gavin Miller (see www.snakerobots.com) has

T. Gomi (Ed.): ER 2001, LNCS 2217, pp. 1–37, 2001.

built robots to investigate and demonstrate several of the known snake 'gaits'. Figure 1 shows his robot prototype S5 executing the classic serpentine locomotion well known to every schoolchild. Heavy-bodied snakes such as pythons tend to use a different method — rectilinear locomotion — in which the body lies in a straight line, and waves of contraction run down the scales along the length of the snake to move it forward like a caterpillar. (Contrary to popular belief, the ribs of snakes are rigidly fixed to the vertebrae. This makes them especially interesting for roboticists because it means that the segments of a snake robot can be rigid.) Tree snakes and some other species use a third method — concertina motion — in which the front part of the snake is lifted and stretched forward, and is then put down and used to drag the rest of the snake forward. Finally, in environments in which there is very little grip available from the substrate — for example, in loose sand — a fourth method, sidewinder motion, can be used. In this, the head and front part of the body are raised and thrown sideways; when they hit the ground, the following part of the body is half-thrown, half-pulled across, and the snake progresses sideways at high speed and with little slippage. Each of these methods represents a good engineering solution to a specific problem, well worth further investigation by the engineer.

A related activity is illustrated by the work of Robert Full (see these proceedings). Rather than simply copying a particular animal implementation, a deep understanding of the principles behind a diversity of animals' solutions to

Fig. 1. Gavin Miller's prototype snake robot S5 performing serpentine locomotion. (Copyright 2000, Dr. Gavin Miller of SnakeRobots.com)

a problem (typically a mobility problem) is used to inform the design of a robot, often resulting in designs which would almost certainly not have been arrived at by conventional engineering analysis.

1.2   Nature as a Source of Engineering Methods

A second reason for imitating living things, and one which is close to the theme of this workshop, is seen when what is imitated is not the design itself, but the design technique. Living things were not designed in response to a specification, like most engineering products, but instead evolved in a cumulative and progressive fashion, with random variations being selected, preserved, and developed on the basis of performance alone. Again, the solutions obtained by applying artificial evolution can be very different from those reached by conventional approaches. They are often strikingly original; they may be extremely difficult to analyse and understand; they are often very economical. Evolution can be applied to robot design in a number of ways. The first approaches, exemplified by the early work of Inman Harvey and Dario Floreano (and their respective colleagues) were concerned with evolving neural network controllers for robots. (See these proceedings for their further development of this work.) Before long, evolutionary techniques were also being applied to the design of the robot's sensors and actuators, the general shape of the robot, and the general form of its control system. At present, the purest and most complete application of evolution to robot design is Jordan Pollack's Golem project (see these proceedings), where both the form and the control system of the robot are the result of synthetic evolution acting to produce an effective capacity for mobility. The hope is usually that the use of evolution rather than conventional design will eventually lead to better robot designs; this hope is given extra force by the common experience that designing robots is a really difficult enterprise, mainly because the different aspects of robots are so tightly interconnected that they cannot easily be separated into relatively independent sub-problems as required by conventional engineering design approaches. Indeed, many have argued that the only solution to the problem of complexity in robot design is to use evolutionary techniques.

There is another natural method frequently used to design control systems in robotics: learning. This is not normally as free and unconstrained as evolution, being limited to producing parametric variations on an architectural theme specified by the designer, but is nevertheless very effective in both natural and artificial systems. Particularly when implemented in the form of neural networks, learning techniques are often able to produce control systems which perform better or are more economical than conventionally designed systems. Modern approaches often combine learning and evolutionary techniques. For example, Dario Floreano's recent work, in which evolution is also allowed to act on the parameters specifying the learning rules of a neural network, has shown that the two techniques used together can easily outperform either used alone on certain tasks.

1.3 Human-Robot Interaction

A third reason for making robots that imitate living things is to enable humans to interact with robots in the same way that they do with other humans and animals. The underlying motivations for this can be diverse. For example, in the toy market there is a long history of dolls and toy animals. Not so very long ago, all such products were passive, and technical progress consisted of making them more realistic by improving appearance and texture, and by providing articulated limbs. After a long period during which minor mechanical and acoustical additions appeared, there was a sudden leap in functionality with the integration of electronics and electric motors. Rather surprisingly to many, these very sophisticated robotic toys do not seem to engage or sustain children's interest for very long. I believe this tells us something very important about the nature and function of toys: play is for developing the child's ability to build and manipulate models of the external world, and toys assist this process by enabling and requiring the child to use his imagination. A toy which is so complete that it leaves no room for imagination will be rejected by the child — yet may continue to fascinate the adult who bought it!

A superficially related but very distinct and important category of artefact is the companion robot, where the interest and rewards for the human come from a relationship involving care, attention-giving, and attention-getting. Even though some companion robots — the outstanding example being the Sony Aibo — might look like toys, they do not function as toys. The challenge for the designer is to discover and implement the behavioural functions that will elicit the desired behaviour and sense of involvement from the human. This is at least as much a psychological task as it is an engineering task. The psychological component is easier than it might appear at first sight, because, as every robot researcher knows, humans have a strong tendency to attribute all sorts of intentional characteristics and higher-level psychological functions to even the most stupid mobile robot. In the light of what comes later, it will be worth spending a little time in discussing the roots of this observation.

The tendency may perhaps be related to animism, originally defined by Piaget [1] as 'the tendency among children to consider things living and conscious'. His early work found that by the age of ten or eleven, most children had developed appropriate categorisations, identifying only plants and animals as being alive. However, further investigation by others (e.g. [2]) has shown that animistic thought appears to persist into adulthood, and that its incidence may be especially high in the elderly [3]. It is only fair to say that doubt has been cast on the reality of this phenomenon in adulthood; the common everyday use, even among scientists, of anthropomorphic terms and metaphorical language can seem like animism if taken literally, and the failure to warn subjects to differentiate between scientific and metaphorical language devalues many studies.

We can gain more insight from the experimental study of the relationship between various object qualities, especially movement, and the attribution by normal subjects of what we might call animate agency, especially cognitive-level animate agency, to the object. In the classic study of Heider and Simmel

[4], the mere movement of simple geometric shapes on a screen in relation to one another proved sufficient to trigger descriptions of the shapes' behaviour in terms of complex motivations. However, since shapes on a screen are clearly not capable of supporting such descriptions in full, the conservative interpretation of the results must surely be in terms of a metaphorical rather than a scientific use of language. With robots, the situation is rather different; the right type of robot could fully support such a description — at least to the extent that an animal could. Nevertheless, we are all very well aware that most of our robots do not have anything corresponding to intentions, and yet naive individuals, including some scientists, will frequently use intentional descriptions, while giving no signs that they intend them only metaphorically.

Some of the most interesting work in this area has been carried out at the Social Responses to Communication Technologies project at Stanford University, where Clifford Nass and Byron Reeves investigate peoples' responses to television, computers, telephones, and other electronic communication devices. Nass and Reeves [5] have identified a phenomenon similar to the attribution of cognitive animate agency: many people treat communications devices as cognitive agents, rather than mere instruments, and relate to them in a social way. They identify the underlying cause as being a lack of evolutionary experience in coping with modern technology: during human evolution, only humans were able to produce the cognitive-level messages and interactions which now also emanate from communications devices, and so the only mechanisms available for responding to these situations are those inescapably associated with humans. Faced with 'non-sentient social responses', we fall victim to the human tendency to confuse what is real with what is erroneously perceived to be real, and behave as if machines are human.

A further type of robot which may have to be designed for interaction with humans is the service robot [6]. A service robot's primary purpose is to perform some service, in the course of which it may have to interface with a human either incidentally (because they share a workspace), or to enable the service to be carried out appropriately. The overwhelming majority of the work carried out into this area involves developing the human-robot interface to allow the person giving instructions to the robot to do so easily and accurately. Some aspects of some interfaces — such as language-based communication — involve the robot mimicking aspects of human behaviour. However, until service robots develop to the point where they must regularly interface with individuals unfamiliar with robots, there is little incentive to designers to consider the possible benefits of making the robots appear more human-like.

1.4 Robots for Understanding Humans

The fourth reason for making robots that imitate living things is perhaps the most important in the long term: we may do so in order to understand how living things, and especially humans, work. Now, what is the meaning of 'work' in this context? I believe there are in fact two meanings. The first is familiar to all of us: it is something like 'produce appropriate behaviour'. This begs the question of what we mean by 'appropriate'. In the context of living beings, the most appropriate action at any time is the one that, in the long term, can be expected to maximise the propagation of the being's genes. Brains are controllers, designed by evolution to do precisely that and nothing else. There is not necessarily any clear parallel for any robot that has yet been designed. However, if the robot has a defined mission, then the most appropriate action may be that which can be expected to maximise the contribution to the success of the mission.

The second meaning exists only in respect of the way in which humans appear to work. The production of behaviour is an external view of how we function. But because humans, and perhaps some other species, are conscious, there is also an internal view — the contents of consciousness. In order to understand fully how humans function, it seems that we must be able to give an account of the external view — behaviour — and also of the internal view — consciousness. And if the two are related, as they appear to be for at least some of the time in humans, we should also be able to give an account of that relationship.

But why should robots play any part in understanding how humans and other animals work — surely that is biology's task? That is of course correct; nevertheless, robotics may be able to make a contribution. When hypotheses about the control of behaviour must be tested, we need to carry out experiments. For ethical or practical reasons, many experiments are impossible to carry out on animals or humans. If we try to resolve this problem by using simulation, we run the risk of getting the wrong answer by inadvertently omitting some critical fact about the real world from the simulation. Real robots offer some defence against this. An excellent example of the use of robots in understanding human behaviour is the Kismet project at MIT (see Rod Brooks' contribution to these proceedings). This is beginning to reveal the details of behaviours underlying human social interactions, and it is difficult to see how any other approach could have delivered so much and in such a convincing manner. A further factor, very appropriate in a workshop on artificial life, is that sometimes science must proceed by synthesis rather than by analysis, especially when emergent phenomena are under investigation. We have all learned a lot about behaviour simply from building and working with robots, and much of this could not have been learned in any other way. A last factor, and much less important, is that robots are excellent demonstration devices: seeing a robot walk, or catch a ball, is a very powerful way of satisfying ourselves and others that we have understood some important aspects of walking or catching balls. If we can build robots with more sophisticated real-world behaviours, then perhaps we may be able to feel that we have some understanding of those behaviours too.

In this paper, I wish to focus on robots designed to increase our understanding of how animals and humans work, and I will deal with both the external and the internal views — with both behaviour and consciousness. There is an Arab proverb: "See the end in the mirror of the beginning". I believe that the end of this robotic enterprise will be the successful construction of a conscious robot; let us see what we can learn from the beginning.

2   Grey Walter and the Imitation of Life

One of the themes of this workshop is 'artificial life'. We think of this as being a very modern idea, first articulated at length by Chris Langton in 1987. In his opening piece 'Artificial Life' in the seminal volume recording the talks at the first workshop in Los Alamos [7], he wrote:

> "Artificial Life is concerned with generating lifelike behaviour. Thus it focuses on the problem of creating behaviour generators." (p5).

But what constitutes lifelike behaviour? Is it something we will recognise when we see it? Or is it possible to define its characteristics? One thing one can certainly say of natural behaviour, in animals at least, is that it achieves something: it has been shaped by evolution to propagate the animal's genes effectively. In fact, in many species, the behaviour selected under various conditions appears to be not merely effective, but optimal: it is the behaviour which, out of all those available to the animal, will make the largest expected contribution to reproductive success. Is it possible to make a list of the different types of behaviour that an animal must possess in order to be able to deal with the various problems it faces in fulfilling its reproductive mission? And if an artificial model animal were able to produce those behaviours appropriately, would it qualify as an example of artificial life? Here is an extract from a paper about just such a model:

"Not in looks, but in action, the model must resemble an animal. Therefore it must have these or some measure of these attributes: exploration, curiosity, free-will in the sense of unpredictability, goal-seeking, self-regulation, avoidance of dilemmas, foresight, memory, learning, forgetting, association of ideas, form recognition, and the elements of social accommodation. Such is life." [8]

This sounds like a plausible, if rather ambitious, programme. If we had to guess at the level of technology necessary to produce this range of behaviour, we would probably imagine something comparable to the Sony Aibo. It comes as something of a shock to realise that the first such model animal (see Figure 2) was built and demonstrated fifty years ago, by the English physiologist Grey Walter. It is perhaps even more shocking to learn that the 'behaviour generators' for most of these behaviours consisted of two vacuum tubes, two sensors, two switches, and a handful of resistors and capacitors. Was this an isolated curiosity, unrelated to our current technologies and concerns? I believe that in fact it sits firmly in the mainstream of modern artificial life robotics, and that much of what we have learned over the past fifteen years could in principle have been learned from a careful study of Grey Walter's work. Let us now examine Grey Walter and his models in more detail.

Grey Walter's work in this area is best viewed in the context of his overall career. Born in Kansas City in 1910 to an Italian mother and a British journalist father, he was educated in England, reading physiology at Cambridge, and then moving to London to begin a research career in neurophysiology. He became interested in electroencephalography, and was the first to observe and

Fig. 2. One of the first model animals, Elsie, returns to the recharging station in her hutch. The notice reads "Machina speculatrix (Testudo). Habitat: W. England. Please do not feed these machines." This picture, from an unknown source, dates from 1949.

identify many of the fundamental phenomena. Although he never formally studied engineering, he had a great talent for seeing opportunities for new electronic equipment, and often played a major role in its design. Before he was thirty, he was appointed Director of Physiology at the newly founded Burden Neurological Institute (BNI) in Bristol, and he remained there for the rest of his working life. After the end of the Second World War, he recruited a talented technical team, many of whom he had met while working on radar during the war years. He made the BNI one of the leading centres for electro-encephalography, a position it maintained even after his death. Among his many important achievements were the identification and naming of delta rhythms (slow rhythms around tumours used for diagnosis and location) and theta rhythms, the discovery of EEG abnormalities in epileptics, the construction of the first on-line EEG frequency analyser, and the discovery of contingent negative variation (also known as the expectancy wave); he also made many other technical breakthroughs in collaboration with a succession of colleagues.

As well as founding the EEG Society, he co-founded both the International Federation of EEG Societies and the EEG Journal. His 174 scientific publications were almost all in the area of electroencephalography and the associated technologies; only a handful are concerned with his model animals. In 1970, he received severe head injuries in a tragic road accident, and spent six weeks in a coma. He made a good physical recovery, but the brain damage from the accident effectively ended his brilliant career, and he achieved nothing more before his retirement in 1975. In 1977 he died of a heart attack.

2.1   Inspirations

What inspired Grey Walter to build his model animals? We can get some idea from his book, "The Living Brain", which was published in 1953 [8]. He writes: "The first notion of constructing a free goal-seeking mechanism goes back to a wartime talk with the psychologist, Kenneth Craik...When he was engaged on a war job for the Government, he came to get the help of our automatic analyser with some very complicated curves he had obtained, curves relating to the aiming errors of air gunners. Goal-seeking missiles were literally much in the air in those days; so, in our minds, were scanning mechanisms. Long before (my) home study was turned into a workshop, the two ideas, goal-seeking and scanning, had combined as the essential mechanical conception of a working model that would behave like a very simple animal."

Although it might be expected that anyone thinking of making a model animal at that time would have been inspired by Norbert Wiener's book, "Cybernetics" [9], we know that this was not the case with Grey Walter. In a letter to Professor Adrian at Cambridge dated 12th June 1947, he wrote: "We had a visit yesterday from a Professor Wiener, from Boston. I met him over there last winter and find his views somewhat difficult to absorb, but he represents quite a large group in the States, including McCulloch and Rosenblueth. These people are thinking on very much the same lines as Kenneth Craik did, but with much less sparkle and humour." In fact, Grey Walter was in close touch with an extensive group of British scientists who formed an independent tradition of cybernetic thought within England. Many of them came together later in the decade to form a discussion group for such ideas, the little-known but highly influential Ratio Club (named by Grey Walter); its members included Alan Turing, I.J.Good, Ross Ashby, and Horace Barlow.

An idea that had interested Grey Walter for some time, and that he probably first heard from Craik, was that a control system operating in a complex world would have to be able to represent that complexity internally. In a revised form this became Ashby's well-known Law of Requisite Variety. It was clear that the brain itself, with its vast number of neurons, was enormously complex, and so it was easy to assume that it was capable of representing the external world adequately because of its own complexity. From the beginning, Grey Walter recognised the sheer impossibility of building a device with anything like the number of functional units that the brain appeared to possess, and so he began to wonder if there was any way in which internal complexity could be achieved using a relatively small number of components. By 1948 he thought he had found a way. In a radio talk on the brain in that year, he spoke as follows:

> "I am going to develop the hypothesis that the functional interconnection between brain cells bears some relation to the processes of thought and consciousness, and my first premise is that the variety of permutation in these connections is at least as great as the diversity and complexity which we are subjectively aware of in our own minds and objectively assume in the minds of others. The general idea behind this hypothesis is that the function of the brain is to make a working model of external reality, and it is clearly of the first importance to establish at the outset that there are as many working parts in the model as there are in the full scale pattern." [10]

What lay behind this was that he had realised that the number of different ways of connecting up a number of elements was very much larger than the number of elements, and that it might be possible to devise a system to exploit that. He began by supposing that each different pattern of interconnection could be made to produce a different system response. He worked out that a system of 1,000 elements could be interconnected in about 10300,000 different ways — a number of responses sufficient, he thought, for any conceivable individual entity "...Even were many millions of permutations excluded as being lethal or ineffective...". But how could different patterns of interconnection give rise to different behaviours? He gained some insight into this problem by reasoning as follows:

> "How many ways of behaviour would be possible for a creature with a brain having only two cells? Behaviour would depend on the activity of one or both of these cells — call them A and B. If (1) neither is active, there would be no action to be observed; if (2) A is active, behaviour of type a would be observed; if (3) B is active, behaviour of type b; if (4) A and B are both active, but independently, there would be behaviour of both types a and b, mixed together; if (5) A is "driving" B, type b would be observed, but subordinate to A; if (6) B is "driving" A, type a would be subordinate to B; if (7) A and B are "driving" each other, behaviour will alternate between type a and type b." (From [8])

It was against this background that Grey Walter designed and built his model animal, Machina speculatrix, commonly known as the "tortoise". In place of nerve cells, it used vacuum tubes. More importantly, in view of the last paragraph, it used only two of them. And most significantly, it used relays to vary the connectivity within the system automatically as a function of environmental circumstances and history.

2.2   The Tortoises

We can date the development of the first tortoises quite accurately. The late Nicolas Walter, Grey Walter's son, remembered discussing his father's work with him in the spring of 1948, and was certain that no mention was made of any model animals. 21 months later, a newspaper article of December 1949 carried pictures of Elmer and Elsie, the first two tortoises, along with a complete and accurate description of their behavioural repertoire. Elmer, 'the prototype', was reported to have been built 'more than a year ago'. They were built in the 'backroom laboratory' of his house by Grey Walter, 'helped by his wife Vivian', a professional colleague for many years.

Elmer and Elsie (ELectroMEchanical Robots, Light Sensitive with Internal and External stability) were the subjects of a famous 1950 Scientific American article, entitled "An Imitation of Life" [11]. Amid the text and illustrations describing their construction and behaviour are phrases strikingly reminiscent of modern Artificial Life: "...synthetic animal...synthetic life...the scientific imitation of life..." Unfortunately, the general style of writing is rather literary and indefinite, and it is not possible to work out exactly how the robots were constructed, or how they produced the reported behaviour. The illustrations are simply rough line drawings with sketches of the trajectories followed by the robots, and with very brief captions. In "The Living Brain" there is a chapter about the tortoises and their behaviour, but it contains no illustrations except for a single photograph. An appendix to the book gives some information about the design of the tortoises, including a 'typical' circuit diagram and an explanation of its electrical function. Although it is possible, by using all three sources of information, to see how the trajectories in the drawings might have been produced, it is extremely difficult, and nowhere is there any material that would nowadays count as evidence. How can we now assess the ability of Grey Walter's model animals to produce lifelike behaviour?

At the University of the West of England, we have attacked this problem with some success. With assistance from the British Broadcasting Corporation, we have found a short newsreel film of Elmer and Elsie, probably from 1949. Thanks to present and former staff at the Burden Neurological Institute, we have found many original documents, illustrations (e.g. Figure 3, and photographs, some of which appear to be the originals from which the sketches in the Scientific American article were prepared. We have found unpublished writings by Grey Walter giving insight into the tortoises [12], and also into Grey Walter's thought processes. We have discovered and restored to working order the only functionally complete surviving tortoise — from a batch built in 1951 by Mr. W.J. 'Bunny' Warren, one of the talented technical team recruited by Grey Walter in 1945. (This tortoise — see Figure 4 — is now on display in the Science Museum in London. A second incomplete and non-functional survivor from this batch is currently in display in the MIT museum.) Finally, we have constructed two replica tortoises based on the rediscovered example, and have thus been able to study both single and multiple tortoise behaviour.

2.3   How the Tortoises Worked

The physical construction of the tortoises is easily understood from Figures 3 and 4. The single driving wheel is mounted between forks attached to a vertical

Fig. 3. A labelled illustration from the files of the Burden Neurological Institute, showing a tortoise (Elsie) with its shell removed. The shell was suspended from the central pillar, and operated the touch contact when displaced.

shaft, on top of which is fixed a photoelectric cell. The shaft can be rotated, in one direction only, by the steering motor; as the shaft rotates, the photoelectric cell and the forks holding the driving wheel rotate with it. When both the driving and steering motors are on, the driving wheel traces out a cycloidal path, and the whole tortoise moves sideways in a looping progressive trajectory. The shroud around the photoelectric cell, which is aligned with the driving wheel, limits its sensitivity to sources of light in line with the current orientation of the wheel. The shell, which is suspended from the top of the tortoise, as can be seen in Figure 4, is deflected whenever it comes into contact with an obstacle, and operates a touch contact until it returns to the resting position. At the front of the tortoise is a 'headlight'; this is lit whenever the steering motor is active.

The electrical circuitry of the tortoises is extremely economical and elegant. The circuit diagram, shown in Figure 5, shows all components. However, the

Fig. 4. The only surviving tortoise in working order, rediscovered in 1995, and now on display in the Science Museum, London. This example was from a batch of six built in 1951.

principles of operation can probably best be understood with the aid of the function diagram in Figure 6. The light falling on the photoelectric cell generates a current which is fed to the first vacuum tube and amplified. The amplified signal is then passed to a relay capable of switching either the turn or drive motor to half-speed, and also goes on to a further stage of amplification by the second vacuum tube. The signal from the second vacuum tube is passed to a second relay capable of switching either the turn or drive motor to full speed. As the light intensity at the photocell increases, the currents to the turn and drive motors vary as follows:

1. Light intensity LOW: the output from the first amplifier is LOW, and so its relay switches the Drive motor to half speed. The output from the second amplifier is also LOW, and its relay switches the Turn motor to full speed. The tortoise moves slowly sideways in a cycloidal path, scanning the room rapidly with the rotating photoelectric cell.
2. Light intensity MEDIUM: the output from the first amplifier is still too LOW to switch the relay over, so the Drive motor is still powered at half speed.

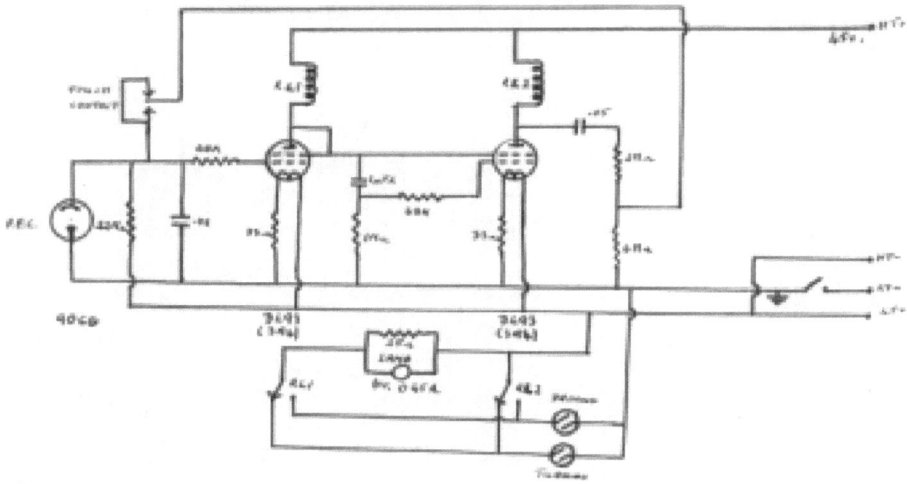

Fig. 5. The circuit diagram for the 1951 batch of tortoises, preserved in the Burden Neurological Institute archives.

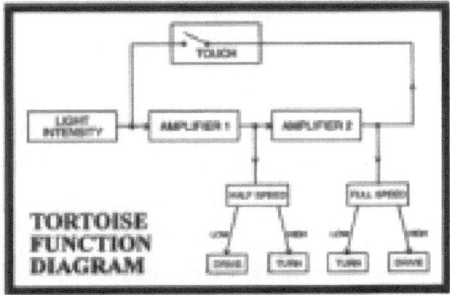

Fig. 6. The tortoise function diagram shows how the Drive and Turn motor currents are controlled by the light (photocell) and touch inputs.

However, after the second stage of amplification, the output becomes HIGH and the second relay switches full power from the Turn motor to the Drive motor. The net effect is that the Turn motor is not powered at all, and the Drive motor operates at full speed. The effect on the tortoise is dramatic: it stops the slow sideways progress, and moves rapidly in the direction in which the photocell and driving wheel are pointing. But since the driving wheel will usually be at an angle to the axis of the tortoise, it will not move in a straight line, but in an arc; since this movement will change the orientation of the photocell, the light source will be lost, and the Turn motor will start up again. One of two things can happen next. If the arc was in the opposite direction to the rotation of the photocell, then the restarting of the Turn motor will cause the photocell to encounter the light again, and so the Turn motor will again be stopped. The result of this is usually a relatively long, smooth, and straight movement towards the light source, sustained for as long as the direction of the arc of movement relative to the photocell rotation does not change. However, if and when the arc of movement is in the same direction as the photocell rotation, the loss of the light source and the restarting of the Turn motor will not lead to the immediate recapture of the light source, and the photocell will have to through a complete rotation before the light source is again aligned with the shroud.

3. Light intensity BRIGHT. In this condition, the outputs from both the first and second amplifier stages are HIGH, and so the Turn motor is powered at half speed and the drive motor at full speed. This produces a movement similar to that produced by LOW light intensity, but on a larger scale. Because the Turn motor is active, the light source is lost almost immediately, and the tortoise reverts to the LOW input mode.

4. TOUCH. If at any time the shell is deflected and the touch switch is operated, the output from the second amplifier is connected back to the input of the first stage. This positive feedback turns the circuit into a low-frequency oscillator, with the outputs from the amplifiers both changing between HIGH and LOW every second or so; when this happens, the circuit is unaffected by any input from the photocell. The oscillation produces an alternation between the outputs corresponding to LOW and BRIGHT light intensity, which persists for slightly longer than the activation of the touch switch because of hysteresis in the oscillator. The effect on the tortoise's behaviour is to produce a combined pushing and turning against the obstacle. If the obstacle is relatively heavy, the tortoise will eventually turn far enough to free itself; if the obstacle is light, the tortoise may push the obstacle out of the way.

We can be sure that the original tortoises produced these basic behaviour patterns, as Grey Walter termed them, because of the preservation of a remarkable series of photographs in the Burden Neurological Institute archives. At some time during 1949 or 1950, at one or more photographic sessions at Grey Walter's house, photographs of trajectories were produced by fixing candles on the tortoises' backs and using long exposures. Figure 8 shows both Elmer and Elsie. Elmer was the first prototype, and was less reliable than Elsie; most of the records feature Elsie alone. In this photograph, several interesting behaviours can be seen. Elmer and Elsie are released at equal distances from the brightly-lit hutch. After an encounter, probably stimulated by the attraction to one another's candles, they separate. At this time, Elsie's bright headlamp clearly shows the sideways cycloidal movement of the front of the tortoise; the candle shows the backwards-and-forwards movements of the rear end. At one point, Elsie's photocell becomes aligned with the light in the hutch when the intensity is sufficient to trigger the MEDIUM light behaviour. Because the orientation of the photocell produces an arc opposing the direction of rotation of the photocell, Elsie produces a long straight movement towards the light source. Note the disappearance of the headlamp trace as the Turn motor is turned off. Closer to the hutch, the alignment is lost, and Elsie reverts to the cycloidal movement.

We have confirmed the existence of these behaviour patterns on the surviving tortoise, and on the replicas. However, they are of little interest in themselves. What is important is the ways in which they can give rise to behaviour that deserves to be called 'lifelike'. Do the photographs support Grey Walter's claims?

2.4   Lifelike Behaviour?

Let us compare Grey Walter's descriptions of the tortoises' behaviour with the evidence from some of the photographs. In "The Living Brain" [8], he identifies several behaviours which he claims to be lifelike, or characteristic of living beings:

Speculation. Walter claims that 'A typical animal propensity is to explore the environment rather than to wait passively for something to happen'. This is implemented by the LOW light behaviour — the cycloidal movement moves the tortoise around the environment, and the rotation of the photocell produces a constant scanning effect. This is seen clearly in some of Elsie's movements in Figure 7.

Positive tropism. As Walter notes, 'The one positive tropism of M. speculatrix is exhibited by its movements towards lights of moderate intensity'. Again, this is visible in Elsie's movement in Figure 7, when she stops the cycloidal movement and heads in a straight line towards the brightly-lit hutch.

Negative tropism. Walter writes, 'Certain perceptible variables, such as very bright lights, material obstacles and steep gradients, are repellent to M. speculatrix; in other words, it shows negative tropism towards these stimuli'. He is therefore claiming that both the BRIGHT light behaviour and the TOUCH behaviour are examples of negative tropisms. I believe he is mistaken on both counts. The BRIGHT light behaviour is simply a transformed version of the sideways cycloidal motion seen in the LOW light behaviour; as soon as the rotation of the photocell causes the source of illumination to be lost, the LOW light behaviour reappears. The tortoise is not exactly repelled by strong light, but moves sideways in response to it — essentially the same neutral behaviour as is manifested towards LOW light. Since the tortoise will have approached the light by moving forwards along an arc, moving sideways from that orientation will tend to carry it away from the light. In the TOUCH behaviour, the tortoise turns and pushes. If the obstacle is light enough to be moved by a strong push, it will be displaced as the tortoise pushes forwards — hardly a negative tropism. If the obstacle is heavy, then the tortoise either stalls its motor or spins its wheels until it has turned far enough to move; this is no tropism, but simply the effects of physical constraints.

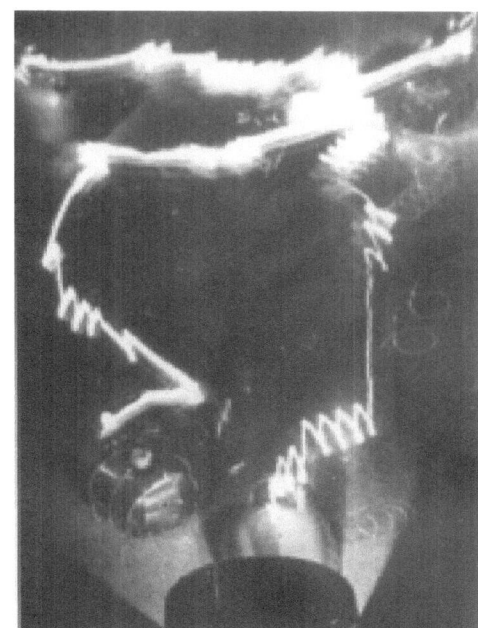

Fig. 7. In this time-exposure photograph, Elmer and Elsie, carrying candles, race for their hutch. Elsie, with the smooth shell, has a bright headlamp that clearly shows the sideways cycloidal movement pattern under low light conditions.

Discernment. By this, Walter means, 'Distinction between effective and ineffective behaviour. When the machine is moving towards an attractive light and meets an obstacle...the induction of internal oscillation does not merely provide a means of escape — it also eliminates the attractiveness of the light, which has no interest for the machine until after the obstacle has been dealt with.' This is surely a case of either over-interpretation or the use of metaphorical language. However, one of the photographs, Figure 8, appears to show this effect.

Figure 9 shows a similar trajectory, but in this case the obstacle is high enough to block the light. After a rather messy collision with the obstacle, and some exploration, the light source becomes visible, and the robot is able to make a clean approach.

Optima. According to Walter, one property of living beings is 'A tendency to seek conditions with moderate and most favourable properties, rather than maxima'. Since the intensity of the photocell signal from a given lamp increases as the tortoise gets closer to the lamp, a tortoise approaching a sufficiently bright light will at some distance switch from the MEDIUM light behaviour (attraction) to the BRIGHT light behaviour (apparent repulsion). This can be seen in Figures 8 and 9. As the tortoise circles round the light, sections of zig-

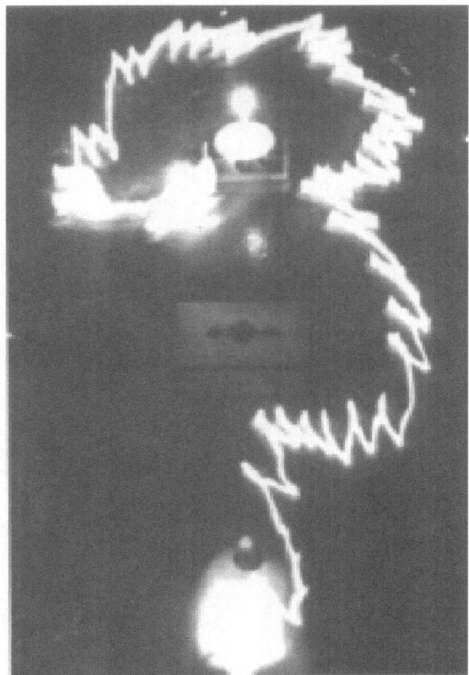

Fig. 8. 'Discernment'. In this time-exposure photograph, Elsie starts at the bottom and moves towards the light. When she encounters the low stool, the TOUCH behaviour interrupts her forward progress several times until she is clear of the obstacle.

zag sideways movement taking the robot away from the light alternate with straighter sections (the attraction behaviour) bringing it back.

Walter also identified a potential problem with tropisms — what he called 'the dilemma of Buridan's ass', a mythical beast that starved to death because it was equidistant from two piles of hay. He notes, 'If placed equidistant from two equal lights, M. speculatrix will visit first one and then the other'. Figure 10 shows a clear example of this. Note the long straight section of MEDIUM light behaviour that enables the machine to make the transit from one light to the other.

Self-recognition and Mutual Recognition. The headlamps of the tortoises are lit whenever the Turn motor is active. Walter originally installed this feature so that he could know the state of the Turn motor without having to infer it from the behaviour. In the early tortoises, the lamps were bright enough to stimulate the photocell when reflected from a mirror, producing the MEDIUM light behaviour. Since the Turn motor is unpowered in this behaviour, the lamp is immediately turned off, the reflection disappears, the LOW light behaviour returns, and the headlamp is again lit. Walter described the consequences: 'The

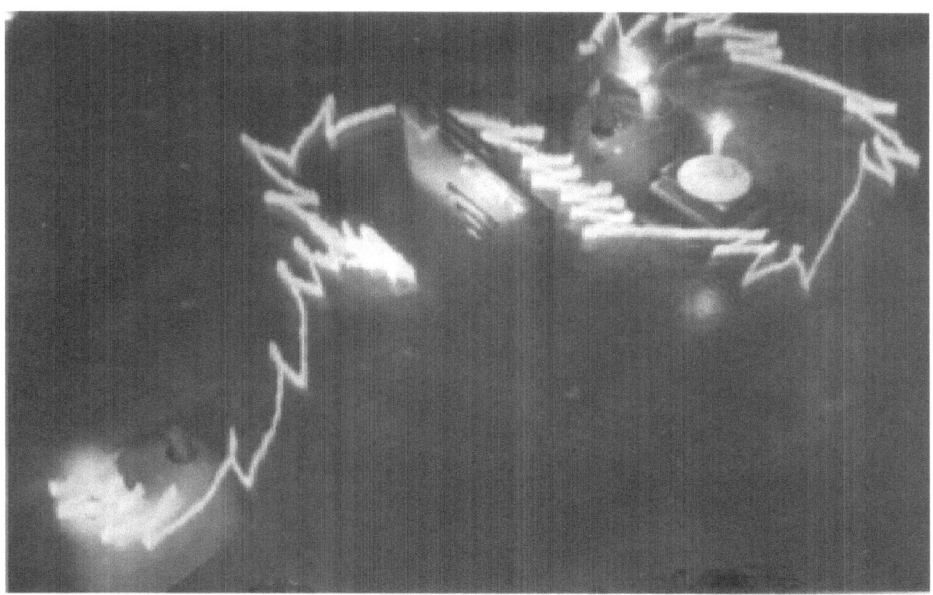

Fig. 9. In this experiment, Elsie encounters an obstacle high enough to obscure the light. She wanders until the light comes into view, and then successfully approaches it.

creature therefore lingers before a mirror, twittering and jigging like a clumsy narcissus. The behaviour of a creature thus engaged with its own reflection is quite specific, and on a purely empirical basis, if it were observed in an animal, might be accepted as evidence of some degree of self-awareness'. Two tortoises together could also attract one another via the headlamps. Walter described the result as follows: 'Two creatures of this type meeting face to face are affected in a similar but again distinctive manner. Each, attracted by the light the other carries, extinguishes its own source of attraction, so the two systems become involved in a mutual oscillation, leading finally to a stately retreat'. Even then, it was clear that Walter was choosing his words extremely carefully to avoid any accusations of over-interpretation while inviting his readers to do just that. His statements would of course be dismissed out of hand by most modern biologists, but it would still be interesting to examine the subjective impressions caused by witnessing the behaviour.

Unfortunately no materials showing 'self-recognition' (known popularly as 'the mirror dance') or 'mutual recognition' have come to light. Interestingly, there is one surviving image from the time exposure sessions showing a tortoise in front of a mirror (Figure 11). However, the pattern of movement is very different from that sketched in [11]; it is not 'twittering and jigging', but shows a regular alternation between moving forwards and backwards. It is most probably due to the reflection of the candle in the mirror, and may consist of successive

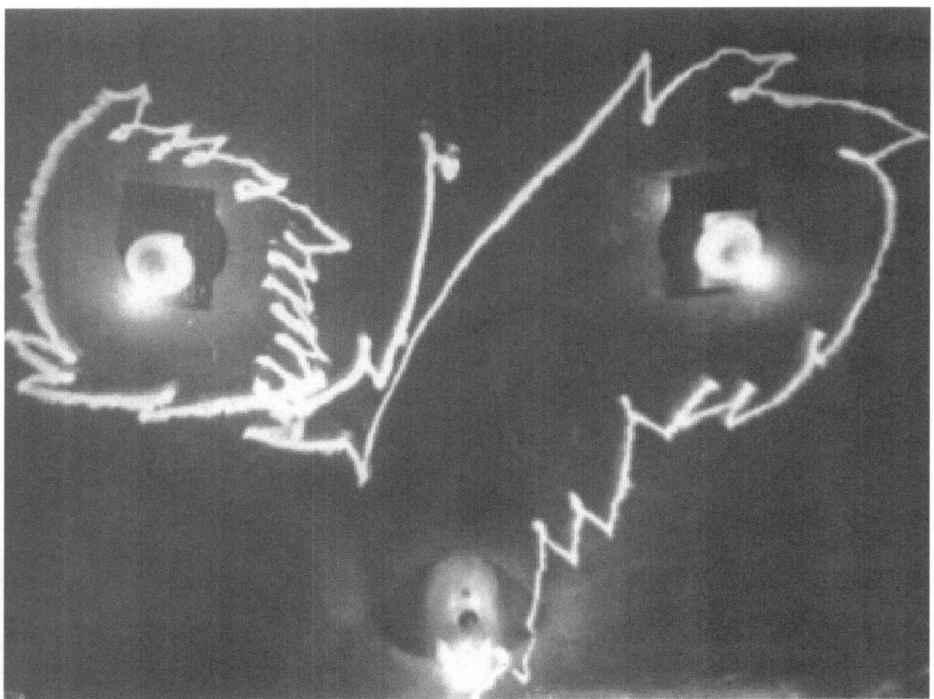

Fig. 10. Presented with two equidistant light sources, Elsie moves first to one, and then to the other.

approaches to the mirror caused by the MEDIUM light behaviour, and retreats caused by the TOUCH behaviour.

Internal Stability. By accident, rather than design, the sensitivity to light of the tortoise circuitry decreased as the battery voltage declined with use. This meant that the brightness threshold causing the transition from the MEDIUM light behaviour (attraction) to the BRIGHT light behaviour (indifference or repulsion) would rise progressively during an experiment, and the tortoise would approach progressively closer to bright lights before turning away. Grey Walter took ingenious advantage of this by fitting a recharging station inside the hutch (see Figure 2) together with a very bright light. At the beginning of a session, the tortoise would move towards the hutch, but would turn away as soon as the BRIGHT light behaviour switched in. However, when the battery became sufficiently exhausted, the tortoise would continue its approach and enter the hutch. As it moved forward, it would encounter an arrangement of switches and contacts which would disable any motion until the batteries were recharged, and then release the robot again. In various passages Grey Walter invites the reader to think of electricity as being the robots' food, and of the recharging

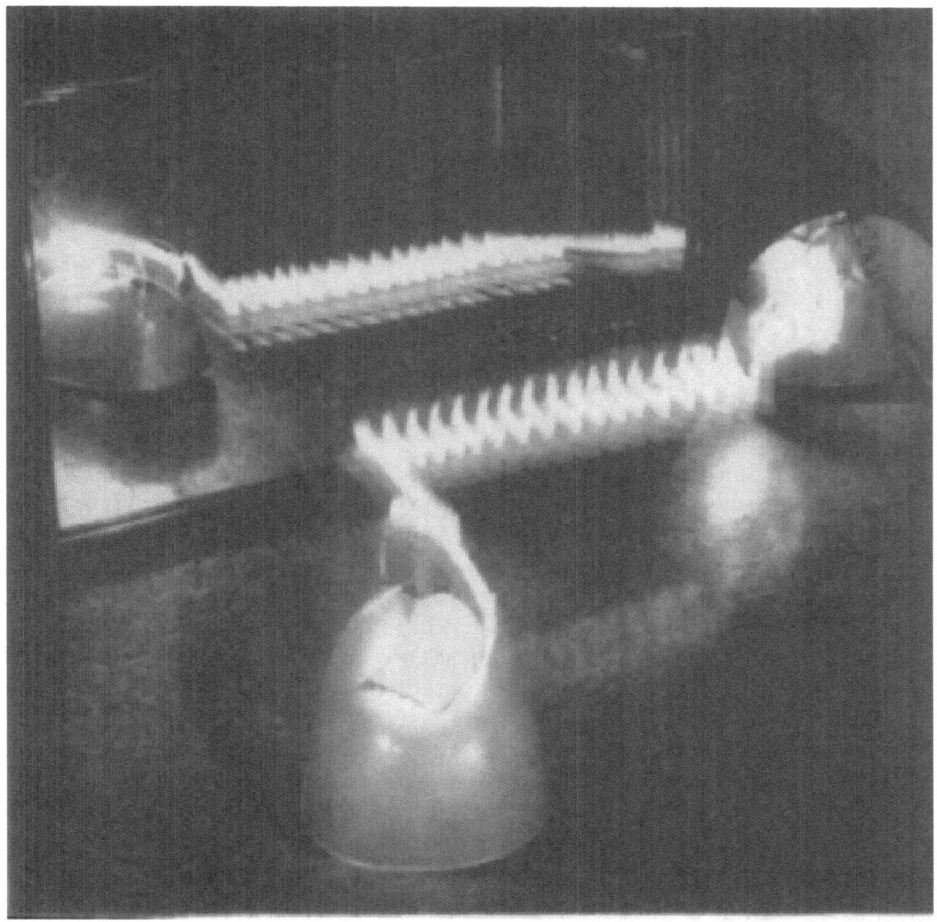

Fig. 11. Elsie performs in front of a mirror, but the pattern of movements does not correspond to descriptions of 'the mirror dance', and is probably induced by the candle she is carrying. Is the candle mounted on top of the photocell? The illustration, found in the Burden Neurological Institute archives, is from an unknown source.

scheme as being the equivalent of eating only when hungry. Unfortunately there is no surviving evidence of this process in action, but many of the details of the recharging station are visible in the BBC film.

Sequences of Behaviour. In an unpublished manuscript found in the archives of the Burden Neurological Institute in 1995, (eventually published in [13]) Grey Walter demonstrated his understanding of how the four basic behaviour patterns (see Figure 12) could be stitched together in apparently useful sequences by the changes in sensory inputs produced by the robots' behaviour, moderated

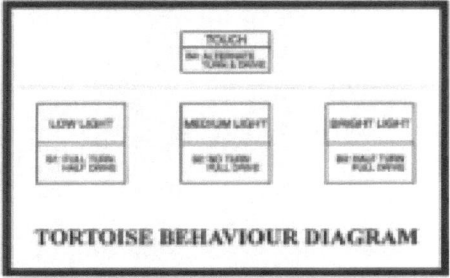

Fig. 12. The tortoise behaviour diagram summarises the motor outputs associated with the photocell and touch inputs.

by some dominance relationships ('prepotency') between the behaviours. This is one of the fundamental techniques in what has become known as behaviour based robotics. He also noted how the tortoises' behaviour in a modifiable environment could produce apparently useful changes in the environment: 'If there are a number of light low obstacles that can be moved easily over the floor and over which the model can see an attractive light, it will find its way between them, and in doing so will butt them aside. As it finds its way toward the light and then veers away from it and wanders about it will gradually clear the obstacles away and sometimes seems to arrange them neatly against the wall. This tidy behaviour looks very sensible but is an example of how apparently refined attitudes can develop from the interaction of elementary reflex functions.' (Quoted in [13])

Figure 13 shows the set-up for such an experiment. Unfortunately there are no accounts or images of this work other than those presented here. It is clearly a forerunner of the section of modern robotics dealing with the use of stigmergy — using environmental modifications to change the future behaviour patterns of robots.

Learning and CORA. In his original list of life-like characteristics, Grey Walter included 'foresight, memory, learning, forgetting, association of ideas, form recognition'. None of these appear to be present in the tortoises. In fact they are introduced by another electronic device (CORA — the COnditioned Reflex Analogue — see Figure 14) originally developed to illustrate some of his ideas about Pavlovian conditioning. Although some of his papers give the distinct impression that a version of CORA had been integrated into a tortoise, none of Grey Walter's colleagues at the time, including the technical staff who would have had to build the device, have any recollection that this was ever done. However, he certainly performed experiments with CORA connected to a tortoise — Figure 15 shows a mock-up of such an arrangement. The tortoise in the London Science Museum, which was Grey Walter's 'personal' tortoise, still retains the connections made to its circuitry to form the interface to CORA.

Fig. 13. Grey Walter with a tortoise from the second batch (with an unusual opaque shell) carrying out an experiment on modifying the environment. The boxes are heavy enough to trigger the tortoise's TOUCH behaviour, but light enough to be pushed. (From the Burden Neurological Institute archives)

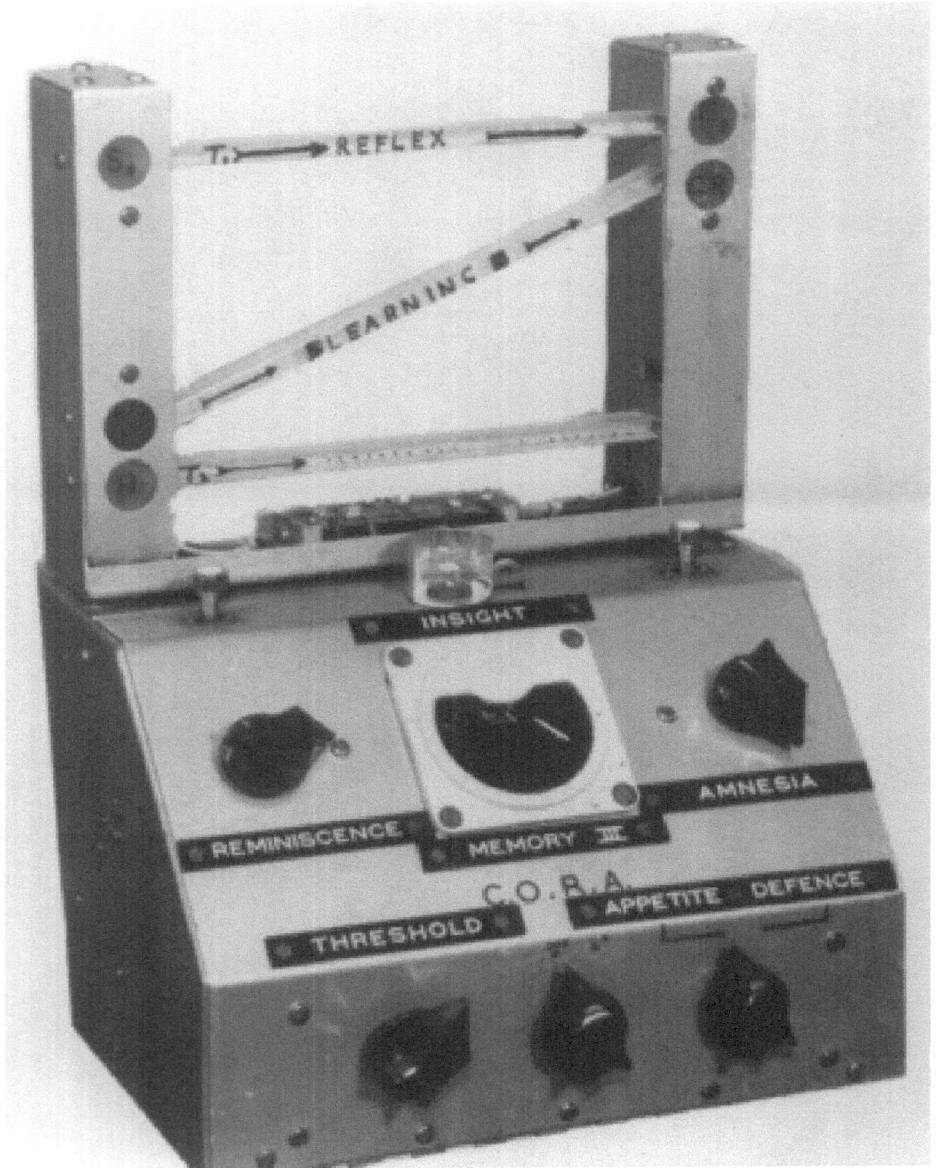

Fig. 14. CORA, the Conditioned Reflex Analogue, currently on display in the Science Museum, London.

Fig. 15. A very serious-looking Grey Walter pretending to carry out an experiment with CORA, a tortoise, and a torch. From the archives of the Burden Neurological Institute.

As far as any scientific merit is concerned, CORA has not worn nearly as well as the tortoises. It is a very ingenious and economical design, but it does no more than accurately implement the straightforward ideas of conditioning current in the 1950s. When CORA was in standalone mode, the progress of conditioning was shown by various lights on the superstructure. Almost the sole virtue of connecting CORA to a tortoise seems to have been pedagogic. No images or films survive showing CORA connected to a moving tortoise, and it seems unlikely that this was ever attempted on a regular basis.

Unfortunately, the restraint that Grey Walter often showed when talking (or at least writing) about the tortoises seemed to desert him where CORA was concerned. In [11], speaking of what appears to be the combination of a tortoise and CORA, he claims, 'One intriguing effect in these higher synthetic forms of life is that as soon as two receptors and a learning circuit are provided, the possibility of a conflict neurosis immediately appears. In difficult situations the creature sulks or becomes wildly agitated and can be cured only by rest or shock — the two favourite stratagems of the psychiatrist'. In spite of these excesses of interpretation, he did recognise that learning was characteristic of living beings, and he did connect the tortoises to an analogue of what was thought at the time to be the basis of much learning. But it is difficult to find any way in which the conjunction made either the tortoises or CORA more lifelike.

What conclusions can we come to about Grey Walter's imitations of life? Here are three of the many lessons that can be drawn from his work.

1. Apparently complex and useful external behaviour patterns can be produced by extremely simple internal processes (Langton's behaviour generators) especially when these processes are affected by the sensory consequences of the robot's movement in the environment
2. Apparently complex, useful, and goal-oriented sequences of external behaviour patterns — compound behaviours — can be produced when behaviour patterns change the sensory input from the environment to initiate or modulate subsequent behaviour patterns.
3. Both sophisticated and naive observers have a strong tendency to attribute internal processes and cognitive factors to robots which they demonstrably do not possess.

We have of course learnt these same lessons all over again in the 1980s and 1990s — they have lost none of their power and fascination. The message for those who wish to understand how humans and animals work is clear: we may very well be able to produce convincing behaviour — a lifelike external view that gives the impression of the existence of a complex internal view — if all we do is pursue the development of adequate minimal behaviour generators. This is perfectly acceptable if one is aiming to produce nothing but useful external behaviour — locomotion, floor cleaning, etc. — or the illusion of a complex internal view, as in an effective companion robot. But we have to ask ourselves whether good behaviour is the only relevant factor: perhaps we should investigate the possibility of building behaviour generators that might allow for the development of an internal view? The final section of the paper will explore this possibility.

3   Towards an Internal View

The first point we must accept in our quest for a robot with an internal view of itself and the world is this: we cannot rely on our subjective judgment that a robot has such a view if that judgment is based solely on our external observation of and interaction with the robot. In 1953, Grey Walter observed that the tortoise "...behaves so much like an animal that it has been known to drive a not usually timid lady upstairs to lock herself in her bedroom". We cannot put this lady's behaviour down solely to her lack of experience of such artefacts. Today, the Sony Aibo, MIT's Kismet, and even the communication devices studied by Nass and Reeves are capable of eliciting strong impressions of animate agency — perhaps at a cognitive level — even in sophisticated observers. In a sense, by rejecting this subjective approach, we are rejecting the idea behind the Turing test. This is not the place for a discussion of this particular issue; instead, I want to focus attention of what other approaches are open to us.

An attractive alternative is to examine the internal processes of the robot, and to compare these with the contents of consciousness we observe in ourselves under similar conditions. If there is a good match, and if the relationship between those processes and the observed behaviour of the robot is similar to that between our own experiences and behaviour, then we may be justified in saying that we have a robot with at least a model of an internal view. This may be a good starting point because an enormous amount of experimental and neuroscientific knowledge about consciousness is now available — a very recent state of affairs. For a variety of reasons, disciplines like psychology and neuroscience steered well clear of consciousness until the last decade or so. In 1988, Bernard Baars wrote "...the twentieth century so far has been remarkable for its rejection of the whole topic (of consciousness) as unscientific"[14]. In 1994, Francis Crick was still able to say "The majority of modern psychologists omit any mention of the problem, although much of what they study enters into consciousness. Most modern neuroscientists ignore it". [15] However, the last few years have seen an explosion of interest and activity, along with the appearance of new tools (e.g. brain scanners) allowing previously unimaginable experiments to be performed. The formation in 1993 of the Association for the Scientific Study of Consciousness provided a multidisciplinary focus, and the ASSC annual conferences play a key role in shaping the field.

One example of the way in which this new knowledge is being deployed is Stan Franklin's work on 'conscious software agents'[1]. Without a doubt, the most comprehensive model of conscious information processing is Baars' Global Workspace Theory [14]. Franklin is building software agents with internal structures and processes corresponding the elements of Baars' theory. The intention is not to produce a conscious agent, but rather to flesh out and explore the implications of the Global Workspace model, and to evaluate the engineering advantages of using such an approach in comparison to more conventional software. It would be perfectly possible, and interesting, to extend Franklin's approach to the design of an embodied agent — a robot — and to examine its performance from the same dual perspective. However, if we were interested in producing a conscious robot by this route, there would then be a risk of getting out no more than we had put in: the internal processes of the robot would resemble those of beings with an internal perspective precisely because we had copied them from such beings.

A rather different approach is hinted at in a paper by Dennett entitled 'Consciousness in Human and Robot Minds'[16], in which he very instructively examines many of the issues referred to here. After considering the difficulties of deliberately setting out to build a conscious robot, Dennett concludes:

"A much more interesting tack to explore, in my opinion, is simply to set out to make a robot that is theoretically interesting independent of the philosophical conundrum about whether it is conscious. Such a robot would have to perform

---

[1] http://www.msci.memphis.edu/~franklin/conscious_software.html

a lot of the feats that we have typically associated with consciousness in the past, but we would not need to dwell on that issue from the outset. Maybe we could even learn something interesting about what the truly hard problems are without ever settling any of the issues about consciousness."

He then goes on to describe some of the thinking behind the Cog project at MIT, viewing it as an implementation of this strategy.

As far as consciousness is concerned, there are two potential problems with the approaches of Franklin and Dennett. The first, and least likely to occur, is this: if the artefact in fact turns out to have some or all of the characteristics of consciousness, whatever they may be, then it may be impossible to decide which architectural factors gave rise to which aspects of consciousness. The second is this: if the artefact does not display any aspects of consciousness, there may be no obvious way of continuing the project. Both problems can be avoided by a third approach, one very much in line with the theme of this conference: Evolutionary Robotics.

4   A Quasi-Evolutionary Approach

The strategy is based on a couple of simple observations:

1. The only creatures known to be conscious (humans) and most of those suspected of being conscious (primates) are also known to have the highest intelligence.
   Consciousness may therefore be associated with high intelligence, either as a necessary component or a by-product.
2. All animals, including humans and primates, evolved under the pressure of the same mission (successful transmission of genetic material)

For high intelligence to have evolved, its possession must have given some net advantage to humans and primates in the performance of the common mission. Since all animal missions are the same in principle, the requirement for high intelligence must therefore derive from the detailed circumstances of the mission.

The proposed approach is simply this: to develop or evolve a series of robots, subjecting each to increasingly difficult mission characteristics requiring high intelligence to deal with them, developing or evolving the architecture until the mission can be achieved, and then moving on to the next mission scenario. In parallel with this, a careful watch would be kept for any signs of any of the characteristics of consciousness or the possession of an internal viewpoint. If successful, this progressive incremental strategy should reveal the mission characteristics requiring an internal view, and also the architectural factors enabling that view. Of course, taken at face value this programme exposes us to some of Dennett's observations: "When robot-makers have claimed in the past that in principle they could construct "by hand" a conscious robot, this was a hubristic overstatement...Robot makers, even with the latest high-tech innovations... fall far short of their hubristic goals, now and for the foreseeable future...(W)hen robot enthusiasts proclaim the likelihood that they can simply construct a conscious robot, there is an understandable suspicion that they are simply betraying an infantile grasp of the subtleties of conscious life." In fact our position is more in line with one of his later comments: "...if the best the roboticists can hope for is the creation of some crude, cheesy, second-rate, artificial consciousness, they still win. Still, it is not a foregone conclusion that even this modest goal is reachable." [16]. But like Cog, it seems worth a try.

It is not at all difficult to produce lists of mission characteristics in ascending order of difficulty. For example, consider the single factor of the nature of the environment. Most of our robots exist in fixed and largely empty environments. Progressively add moving objects, objects that can be moved by the robot, objects of different value to the robot, predatory agents, prey agents, cooperative agents, competing agents, and perhaps mate agents, and things become more interesting — and difficult.

One great advantage of this programme is that, like evolution, it can begin with agents showing no signs of consciousness, or even anything approaching intelligence. But two big questions must be answered first. The first is: Exactly how will the robot designs be improved — by evolution, learning, or design? And the second is: Will there be any bias towards using particular types of architectural features believed to underpin intelligence and/or consciousness?

The answer to the first is: Probably by all three. The rule will be something like "Design when you can; learn or evolve when you must", with the proviso that any improvement, however arrived at, can be passed on to later robots.

The answer to the second is: Yes — modelling. In spite of the fact that we know that intelligent behaviour can be produced in the absence of representations of what is currently in the world [17], our experience of the world seems to centre around models, and in particular manipulable models. As Grey Walter said, "The general idea ...is that the function of the brain is to make a working model of external reality"[10]. Richard Dawkins points out in an intriguing passage: "Survival machines that can simulate the future are one jump ahead of survival machines who can only learn in the basis of overt trial and error...The evolution of the capacity to simulate seems to have culminated in subjective consciousness...Perhaps consciousness arises when the brain's simulation of the world becomes so complete that it must include a model of itself".[18]

5   Modelling

Coming back to earth, it is worth considering just what modelling can do. There seem to be four main ways in which modelling can assist an autonomous agent:

1. By assigning novel data to existing models, and so cueing existing actions
2. By detecting anomalies — novelty, errors, change etc.
3. By enabling and improving control
4. By informing decisions, and enabling planning

Even though some forms of modelling have been attacked in the past — and with reason — it is undeniable that the last of these, planning, is what makes humans so intellectually powerful, and planning is simply impossible without some form of modelling. If consciousness is a by-product of high intelligence, then it may well be that it is a by-product of whatever is necessary to support planning.

In order to give ourselves an idea of what the beginning of such a project might look like, and to clarify our thoughts, we have carried out a small pilot study at the California Institute of Technology. Our aim was to find a simple modelling scheme that would allow us to demonstrate the four uses of models on a simple mobile robot in a static environment. We very soon found what we wanted in ARAVQ (Adaptive Resource Allocating Vector Quantization), a technique developed by Linaker and Niklasson [19,20] for analysing extracted abstract sensory flow representations. The problem that interested them was this: It is useful for a robot to store its past sensory 'experiences', but memory limitations usually mean that relatively few experiences can be stored: can sensory flows be encoded economically in a useful form? They observed that the sensory inputs and motor outputs in many parts of a typical robot environment are often relatively stable for long periods, as when following a wall, or traversing a corridor. Using a simple method for clustering sensory data, they enabled the robot to describe such runs using a single 'concept'; long sequences of sensory flow could then be stored economically by labelling the different 'concepts', and recording the number of times each concept was repeated consecutively. Even better, they were able to take advantage of the fact that the 'concepts' were formed directly in the input space, by 'inverting' each instance of a 'concept' to show the positions of environmental features (walls, corridors, corners) that would produce the appropriate sensory data. We refer the reader to their papers [19,20] for a detailed description of their algorithm.

Linaker and Niklasson had used a simulation of the popular Khepera robot, which certainly qualifies as simple from the sensory point of view. We decided to begin with a similar simulation, but to migrate to a real Khepera as soon as possible. Our first exercise was to reimplement their scheme on the latest version of Webots, a flexible sensor-based simulation environment from Cyberbotics S.A. for a range of robots including the Khepera. Figure 16 shows the first experimental environment. The simulated Khepera is released into the environment under the control of a simple wall following routine, and rapidly builds a stable set of 'concepts' corresponding to the obvious features of the environment.

Figure 17 shows how 'concepts' are inverted to yield the positions of environmental features. The figure 17(a) corresponds to a 'concept' observed in corridors: both motors are equally activated, and the two leftmost and the two rightmost of the eight forward infra-red ranging sensors indicate a left and a right wall. For each successive instant at which the 'concept' was recorded as being active, the robot is moved in virtual space by an amount corresponding to the motor activation specified by the 'concept'. In the case of the corridor 'concept', the successive movements trace out the path of the corridor, and the

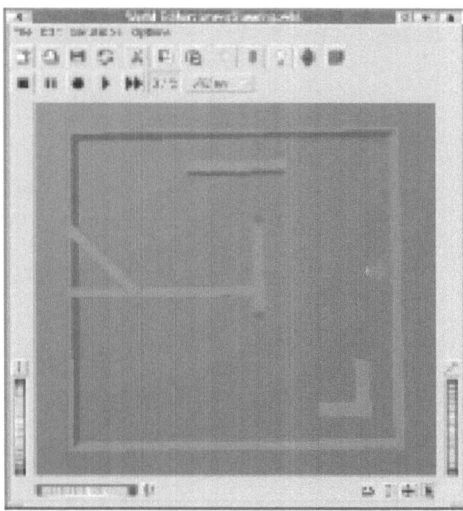

Fig. 16. A screenshot of the environment and Khepera modelled in the Webots simulator. (Copyright Cyberbotics S.A.)

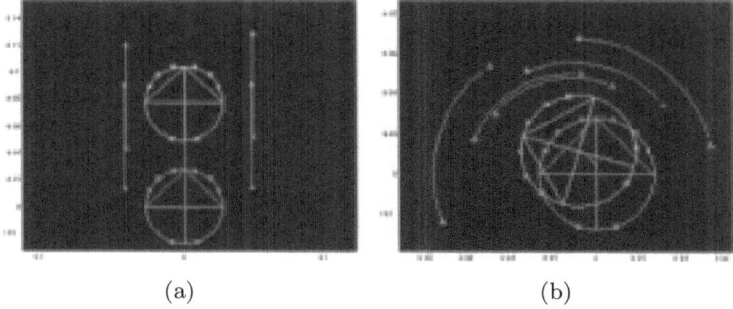

(a)   (b)

Fig. 17. How a map of the environment can be reconstructed from the record of the robot's concept sequence. In the record on the left, the 'corridor' concept is active for two successive time intervals. The model of the Khepera is displaced by an amount corresponding to the concept's motor activation. The sensor readings for the two leftmost and two rightmost sensors give the distance of the reflecting surfaces (walls) at each position; by joining successive corresponding points, we can see part of an environmental map. The record on the right shows a transition involving a small translation and a large rotation, with the derived map. After Linaker and Niklasson, 2000).

sensor range data trace out the left and right walls. The figure 17(b) shows the construction of a corner by the same method. Figure 18 shows the representation of the environment of Figure 16, formed by tracing out the inverted 'concepts' for the appropriate number of steps and in the appropriate order for a typical run. While the global map is distorted - the beginning and end points do not

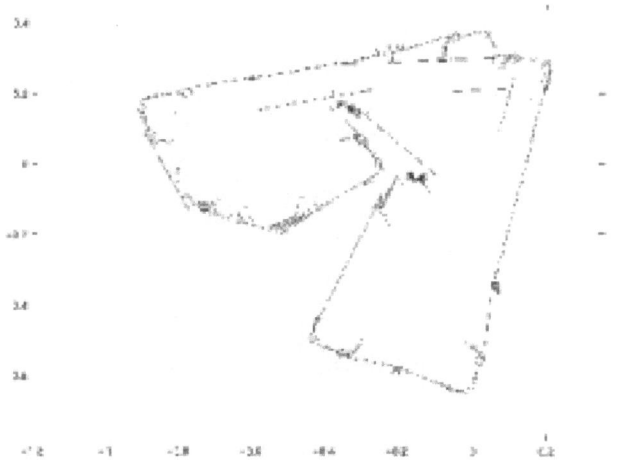

Fig. 18. The environmental map derived from a simulated Khepera in the environment of Figure 16. Although rotational inaccuracies distort the map, it contains good local representations.

coincide — the local representation is excellent. In a sense, this map constitutes the robot's internal model of the environment.

Linaker and Niklasson did not investigate the possible use of 'concepts' by the robot to undertake a task. However, their basic scheme already shows two features of models in general. First, once the robot has established stable 'concepts', it assigns some novel sensory data to those existing 'concepts' on the basis of similarity, or 'expectation'. Second, when new data are very different from any existing 'concepts', it assigns them to a new concept. This corresponds to a kind of anomaly detection.

Since we were interested in how the robot could use the 'concepts', we investigated their possible use for control. The robot had learned the 'concepts' while executing wall following. Could the 'concepts' be used to move the robot? It turned out that a simple method produced apparently good control, enabling the robot to wall-follow round novel environments with more smoothness and accuracy than were evident in the original movement pattern. All that was necessary was to examine the current sensory input, map it to the nearest existing 'concept', and drive the motors with the activations specified by the 'concept'. Figure 19 shows a real Khepera moving round a novel environment under this type of control.

We next tested the ability of a real Khepera to form stable and potentially useful concepts in a real environment. Figure 20 shows the experimental arrangement. There did not seem to be any problems with the transition from

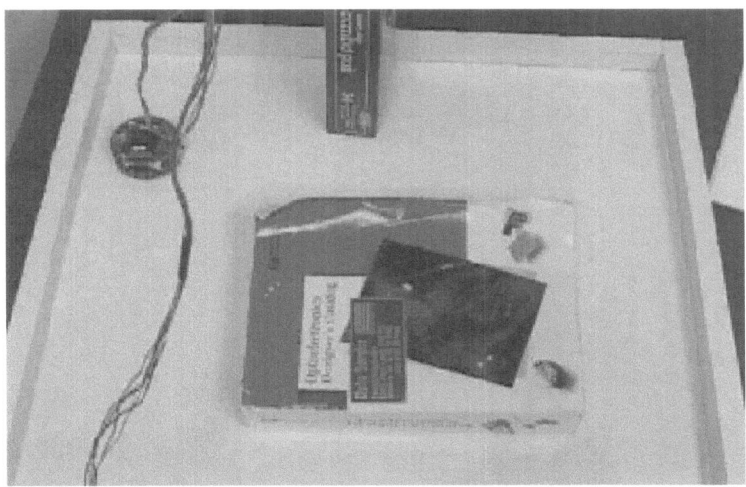

Fig. 19. Using the learned model for control of a real Khepera. The sensor inputs (but not the motor activations) are used to determine the nearest concept, and the motor activations for that concept are sent to the motors. Control appears noticeably superior to the original wall-following behaviour.

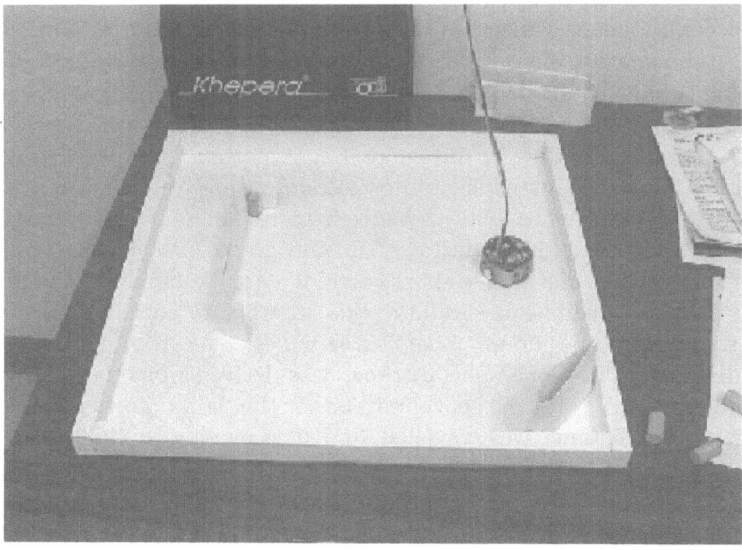

Fig. 20. Learning the map of a new environment using a real Khepera.

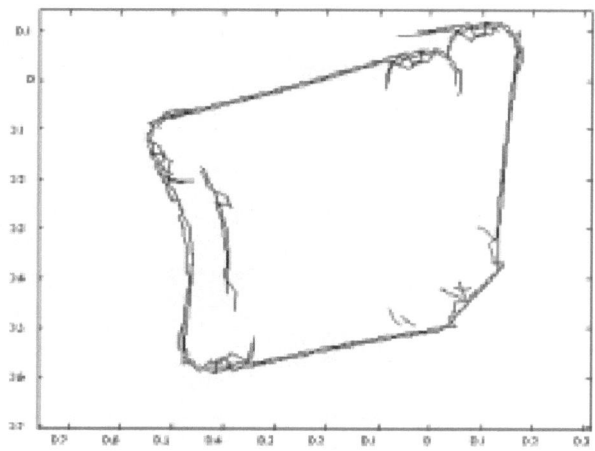

Fig. 21. The environmental map learned in Figure 20.

simulation to reality. Figure 21 shows the reconstruction of the map built by the Khepera in Figure 20. It is qualitatively at least as good as the maps obtained in simulation.

The next step was trivial. Since the robot stores the sequence of 'concepts' derived from a particular fixed environment, it is easy to change the environment to produce a sequence which deviates from the learned norm. This is readily spotted, and constitutes a second type of anomaly detection using a model.

The final challenge in this pilot scheme was to use the robot's internal model to inform some decision. Since all environments were dealt with by performing a right wall-follow, leading to invariant linear paths, there was no clear way of making a choice about which route to follow. However, consider an animal spotted by a predator when some distance from its home: if close enough, it may choose to run for home; if too far away, it may freeze. We declared one part of the environment — a distinctive short corridor — to be home for the robot, and arranged that, at a signal, it should decide whether to continue moving round the environment to the 'home' location, or whether to stop, depending on the current distance from home. This distance was derived from the internal model by stepping forward in virtual space until either the 'home' concept was reached, or the maximum distance was travelled, and then moving to 'home' or stopping, as appropriate.

Even though the individual experiments in this pilot study were extremely simple and obvious — given that we already knew that ARAVQ performed well in simulation, and would probably work on a real Khepera — they do show that it is possible to ask and answer some sensible questions about the use of models in a fixed environment. There is no sign of anything that even the most ardent roboticist could claim as being even a forerunner of consciousness, but we have established certain basic functionalities often associated with consciousness. Although our eventual project will use some form of neurally-inspired controller rather than the purely algorithmic ARAVQ, we can see that the form of the investigation appears to be valid and potentially informative, and so we will continue on this path.

6   The Road Ahead

When will this line of investigation become truly interesting? There is a stage which I believe will be a turning point if it is ever reached. Consider Figure 22. This shows a very conventional view of how a robot might approach tasks in the world — by constructing a world model, updated as necessary, which it can manipulate for a variety of purposes, such as planning or control. For a robot such as a Khepera, operating in a simple environment, such an approach would surely be adequate. Now consider Figure 23. The world model has split into a self model and an environment model, with the self model being able to operate on the environmental model, and also being affected by it. Information generated by the interaction between the two model components somehow makes its way to the executive system. Both the self model and the environmental model receive updates from the external world. Note that the self model does not act directly on the external world, nor perceive it directly. The splitting has produced a system containing an entity — the model self — which seems to be insulated from perception and action in the real world, rather like the conscious self. If such a splitting of the world model can be induced in a robot, it will surely be interesting to examine the characteristics of the various components of the system. However, we cannot say at this stage whether any such splitting can be made to occur, or whether the model self will be of any relevance to the study of consciousness — it is merely an interesting possibility.

What factors might contribute to such a splitting of the world model? There are many possibilities. If the body is very complex, with articulated structures, emotional responses, and so on, then the interrelationships and correlations between the different parts of the body might tend to form a coherent sub-system within the world model, especially since the body will be much less variable that an unconstrained environment. If each robot is evolved from a previous model, then some inherited computational structure may evolve to coordinate the relatively predictable body. Once the environment becomes so complex and the robot

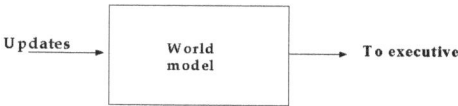

Fig. 22. A conventional unitary world model.

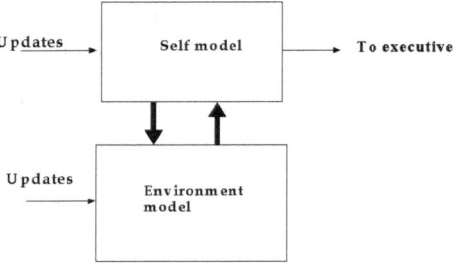

Fig. 23. How the world model may split into a self model and an environment model, with each affecting the other. This raises the intriguing possibility that conscious experience may correspond to the 'experience' of the self model.

so sophisticated that arbitrary multi-step planning becomes worthwhile, then the robot has to predict its own future behaviour accurately, perhaps leading to a segregation of self from the environment. And so on; only the experiments will offer resolution.

Does Grey Walter's success in illuminating the roots of behaviour give us any positive pointers for this enterprise of producing an imitation of consciousness? Possibly. He began with a list of what he thought he would have to demonstrate to achieve his goal. He kept to a strong and thoroughgoing minimalism. He presented his evidence, and invited the reader to form a judgment, rather than claiming anything with the status of a proof. These still seem to be good ideas. It is worth pointing out that the subject of his work — the production of an imitation of life - was at the time at least as fundamentally shocking as a claim to have demonstrated machine consciousness would be now. It is perhaps also worth pointing out that he didn't give up the day job.

Acknowledgments. The present and former staff of the Burden Neurological Institute gave invaluable assistance with tracking down Grey Walter material. The Director, Dr. Stuart Butler, kindly gave permission for the use of archival material. The ARAVQ experiments at Caltech were carried out by Lerone Banks and Nathan Grey, working under the guidance of Rod Goodman.

This work is supported in part by the Center for Neuromorphic Systems Engineering as part of the National Science Foundation Engineering Research Center Program under grant EEC-9402726.

References

1. Piaget J. (1983). Piaget's theory. In Mussen P.H. (Ed.), *Handbook of child psychology*. (Vol 1; pp. 103-128). New York: Wiley
2. Sheehan N. (1980) *Journal of Ageing and Human Development* **12**: 1-13
3. Dennis W. and Mallinger B. (1949) Animism and related tendencies in senescence. *Journal of Gerontology* **4**: 218-221

4. Heider F. and Simmel M. (1944) An experimental study of apparent behaviour. *American Journal of Psychology* **57**: 243-259
5. Reeves B. and Nass C. (1998) *The Media Equation.* Cambridge University Press
6. Engelberger J.F. (1989) *Robotics in Service.* MIT Press, Cambridge, MA
7. Langton C.G. (1988) Artificial Life. In *Artificial Life*, ed. Langton C.G. Vol VI Santa FE Institute Studies in the Sciences of Complexity, Addison Wesley. pp 1-48
8. Walter W.G. (1953) *The Living Brain.* W.W. Norton & Co, New York
9. Wiener N. (1948) *Cybernetics.* Wiley, New York
10. Script of unknown radio talk by Grey Walter, Burden Neurological Institute archives.
11. Walter, W.G. (1950) An Imitation of Life, *Scientific American*, May, 1950, pp 42-45
12. Unpublished notes by Grey Walter (ca 1961): Machina Speculatrix - Notes on Operation. Burden Neurological Institute archives.
13. Holland O. (1996) Grey Walter: the pioneer of real artificial life. In *Artificial Life V*, eds. Langton C.G. and Shimohara K. MIT Press, Cambridge MA
14. Baars B. (1988) *A Cognitive Theory of Consciousness.* Cambridge University Press
15. Crick F. (1994) *The Astonishing Hypothesis: The Scientific search for the Soul.* Scribners, New York
16. Dennett D.C. (1994) Consciousness in Human and Robot Minds. IIAS Symposium on Cognition, Computation and Consciousness. In Ito, et al., eds., *Cognition, Computation and Consciousness*, Oxford University Press
17. Brooks R.A. (1997) Intelligence without Representation. (Revised and extended version) *In Mind Design II*, Ed. Haugeland J. MIT Press. pp 395-420
18. Dawkins R. (1976) *The Selfish Gene.* Oxford University Press
19. Linaker F. and Niklasson L. (2000) Extraction and inversion of abstract sensory flow representations. In *From animals to animats 6: Proceedings of the Sixth International Conference on the Simulation of Adaptive Behavior.* Eds. Meyer J-A, Berthoz A., Floreano D., Roitblat H., and Wilson S.W. MIT Press, Cambridge MA
20. Linaker F. and Niklasson L. (2000) Time series segmentation using an Adaptive Resource Allocating Vector Quantization network based on change detection. *Proceedings of IJCNN 2000*, to appear.

Evolution of Spiking Neural Controllers for Autonomous Vision-Based Robots

Dario Floreano and Claudio Mattiussi

Evolutionary & Adaptive Systems, Institute of Robotics
Swiss Federal Institute of Technology, CH-1015 Lausanne (EPFL)
Dario.Floreano@epfl.ch, Claudio.Mattiussi@epfl.ch

Abstract. We describe a set of preliminary experiments to evolve spiking neural controllers for a vision-based mobile robot. All the evolutionary experiments are carried out on physical robots without human intervention. After discussing how to implement and interface these neurons with a physical robot, we show that evolution finds relatively quickly functional spiking controllers capable of navigating in irregularly textured environments without hitting obstacles using a very simple genetic encoding and fitness function. Neuroethological analysis of the network activity let us understand the functioning of evolved controllers and tell the relative importance of single neurons independently of their observed firing rate. Finally, a number of systematic lesion experiments indicate that evolved spiking controllers are very robust to synaptic strength decay that typically occurs in hardware implementations of spiking circuits.

1 Spiking Neural Circuits

The great majority of biological neurons communicate by sending pulses along the axons to other neurons. A pulse is a small current charge that occurs when the voltage potential across the membrane of a neuron exceeds its threshold. The pulse is also known as "spike" to indicate its short and transient nature. After emitting a spike, a neuron needs some time to reestablish its electrochemical equilibrium and therefore cannot immediately emit a new spike, no matter how strong its excitatory input is. A typical neuron in the cortex "fires" approximately 10 spikes per second during resting conditions and can emit up to 300 spikes per second in operating conditions. Other neurons can fire more frequently (for example 500 spikes per seconds) clustered in short periods of time ("bursting neurons").

In the field of artificial neural networks we find two different classes of models that differ with respect to the interpretation of the role of spikes. Connectionist models of neural networks [19], by far the most widespread models, assume that what matters in the communication among neurons is the *firing rate* of a neuron. The firing rate is the average quantity of spikes emitted by the neuron over a relatively long time window (for example, over 100 ms). This quantity is represented by the activation level of the neuron. For example, a neuron characterized by a sigmoid activation function, such as the logistic function $f(x) = 1/1 + \exp(-x)$, that gives an output of 0.5 would be equivalent to a spiking neuron that emits approximately half its maximum rate of spikes (150 out of 300 spikes per second, e.g.). Models of pulsed neural networks [14] instead assume that the *firing time*, that is the precise time of emission of a single spike, can transmit important information for the post-synaptic neuron [23]. Therefore, these models use more complex activation functions that simulate the emission and reception of spikes on a very fine timescale.

Spiking circuits have at least two properties that make them interesting candidates for adaptive control of autonomous behavioral robots:

- The intrinsic time-dependent dynamics of neuron activation could detect and exploit more easily (e.g., with simpler circuits or with higher reliability) temporal patterns of sensory-motor events than connectionist neural networks.
- Since the physics of circuits of sub-threshold transistors (i.e., characterized by gate-to-source voltage differences below their threshold voltage) implemented with analog Very Large Scale Integration technology [15] match the properties of spiking neurons, it possible to implement large networks of spiking neurons in tiny and low-power chips [9].

Designing circuits of spiking neurons with a given functionality is still a challenging task and the most successful results obtained so far have focused on the first stages of sensory processing and on simple motor control. For example, Indiveri et al. [8] have developed neuromorphic vision circuits that emulate the interconnections among the neurons in the early layers of an artificial retina in order to extract motion information and a simple form of attentive selection of visual stimuli. These vision circuits have been interfaced with a Koala robot and their output has been used to drive the wheels of the robot in order to follow lines [10]. In another line of work, Lewis et al. have developed an analog VLSI circuit with four spiking neurons capable of controlling a robotic leg and adapting the motor commands using sensory feedback [13]. This neuromorphic circuit consumes less than 1 microwatt of power and takes less than 0.4 square millimeters of chip area.

Despite these interesting implementations, there are not yet methods for developing complex spiking circuits that could display minimally-cognitive functions or learn behavioral abilities through autonomous interaction with the environment. Furthermore, the potentially complex dynamics that a spiking circuit with feedback loops can display allows several alternative functioning modalities. For example, a spiking circuit could use inhibitory neurons to modulate the overall excitation of the network and/or selectively interact with other neurons to inhibit specific actions. Also, the same circuit could use firing rate instead of firing time as the preferred functioning modality (this concept will be discussed in more detail later on in this paper).

Artificial evolution is therefore an interesting method to discover spiking circuits that autonomously develop desired behavioral abilities for robots without imposing constraints on their architecture and functioning modality. In this paper, we describe some initial explorations in the evolution of spiking circuits for

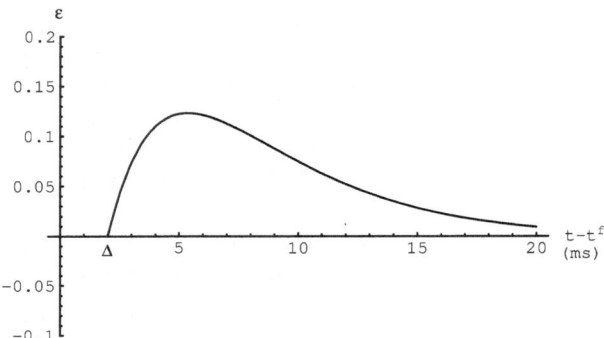

Fig. 1. The function describing the contribution ϵ of a spike from a presynaptic neuron emitted at time t^f. The contribution of the spike begins after some delay Δ (2 ms) due to the traveling time of the spike and eventually decreases its effect as time t flows from the firing time t^f. The synapse time constant τ_s is set to 10 ms and the membrane time constant τ_m is set to 4 ms.

a task of vision-based navigation using a Khepera robot with a linear CMOS vision system. We will then analyze the evolved spiking circuit and discuss the most important lessons learned from these experiments. Finally, we will describe some ideas for future developments in this emerging area of research.

2   The Spike Response Model

The state of a spiking neuron is described by the voltage difference across its membrane, also known as membrane potential v. Incoming spikes can increase or decrease the membrane potential. The neuron emits a spike when the total amount of excitation induced by incoming excitatory and inhibitory spikes exceeds its firing threshold θ. After firing, the membrane potential of the neuron resets its state to a low negative voltage during which it cannot emit a new spike, and gradually returns to its resting potential. This recharging period is called the *refractory period*.

There are several models of spiking neurons that account for these properties with various degrees of detail. In the experiments described in this paper, we have chosen the *Spike Response Model* developed by Gerstner [5]. It has been shown that several other models of spiking neurons, such as the class of Integrate-and-Fire neurons (where the membrane potential of the neuron is immediately reset to its resting value after a spike), represent special cases of the Spike Response Model [6].

In this model, the effect ϵ of an incoming spike on the neuron membrane is a function of the difference $s = t - t^f$ between current time t and the time when the spike was emitted t^f. The properties of the function are determined by *a)* the delay Δ between the generation of a spike at the pre-synaptic neuron

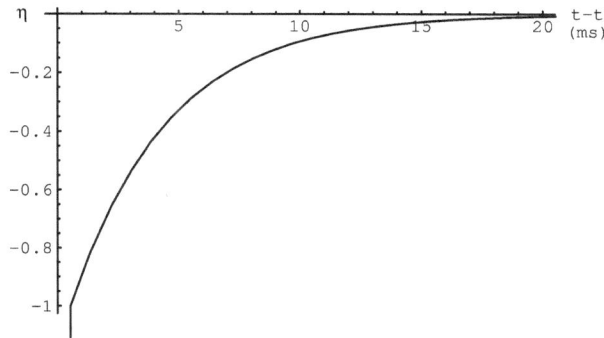

Fig. 2. The function describing the refractory period of the neuron after emission of a spike. Notice the at time 0, the membrane potential is set to a large negative value (larger than 1.0 shown on the y-axis) to prevent another spike. The refractory period is a function of time from last spike and its shape is given by the membrane time constant τ_m (here set to 4).

and the time of arrival at the synapse, *b)* a synaptic time constant τ_s, and *c)* a membrane time constant τ_m. The idea is that a spike emitted by a pre-synaptic neuron takes some time to travel along the axon and, once it has reached the synapse, its contribution on the membrane potential is higher as soon as it arrives, but gradually fades as time goes. A possible function $\epsilon(s)$ describing this behavior, shown in figure 1, is [6]

$$\epsilon(s) = \begin{cases} \exp[-(s-\Delta)/\tau_m](1 - \exp[-(s-\Delta)/\tau_s]) & : & s \geq \Delta \\ 0 & : & s < \Delta \end{cases} \tag{1}$$

In what follows, we will discretize time in small steps of 1 ms each and consider the effects of $\epsilon(s)$ only within a time window of 20 ms. At each time step t, the synaptic contribution to the membrane potential is the sum of all the spikes arriving at the synapse over a 20 ms time window weighted by the corresponding value of the function $\epsilon(s)$ at each time step. Therefore, if at time step t_1 the pre-synaptic neuron has emitted three spikes, respectively at $s = 4, s = 7, s = 15$, the total contribution E_{t_1} computed using equation 1 is

$$E_{t_1} = \epsilon(4) + \epsilon(7) + \epsilon(15) = 0.109945 + 0.112731 + 0.028207 = 0.250883 \tag{2}$$

At the next time step t_2, assuming that there are no new spikes, we simply shift all the previous firing times of each spike by one position (respectively at $s = 5, s = 8, s = 16$) and obtain a new (lower) value $E_{t_2} = 0.2458538$.

Assuming that at time t the neuron membrane is at resting potential v_r, a spike is emitted if $E_t \geq \theta$, that is if the total contribution of the synapse is larger than the threshold. Once the neuron has emitted a spike, its membrane potential is set to a very low value to prevent an immediate second spike and then

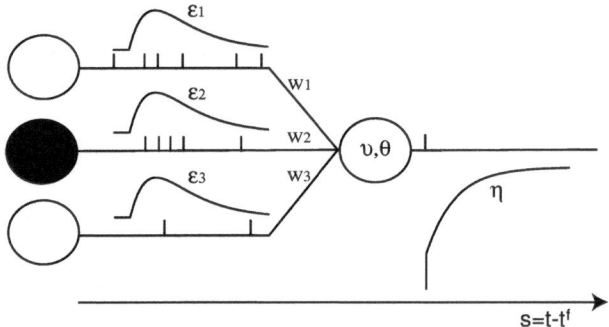

Fig. 3. A network of 4 neurons (white circle= excitatory, black circle=inhibitory) captured at a given time t with the functions ϵ and η superimposed. The postsynaptic neuron has just emitted a spike because the membrane potential v has exceeded its threshold θ.

it gradually recovers to its resting potential. The speed of recovery depends on the membrane time constant τ_m (that we have already seen in the computation of the synaptic contribution above). A possible function $\eta(s)$, shown in figure 2, describing the refractory function [6] is

$$\eta(s) = -\exp[-s/\tau_m] \tag{3}$$

We can now put together the equations describing synaptic contributions and the refractory period to describe the dynamics of a neuron that has several synaptic connections from excitatory and inhibitory neurons (figure 3). Each synaptic connection has a weight w_j whose sign, in this example, is given by the pre-synaptic neuron (positive if the neuron is excitatory, negative if the neuron is inhibitory). The membrane potential of a neuron i at time t is given by

$$v_i(t) = \sum_j w_j^t \sum_f \epsilon_j(s_j) + \sum_f \eta_i(s_i) \tag{4}$$

where $s_n = t - t_n^f$ is the difference between the time t and the time of firing t^f of neuron n. If the membrane potential $v_i(t)$ is equal or larger than the neuron threshold θ_i, the neuron emits a spike and η_i takes a very low value that prevents an immediate new spike. After that, η_i is computed according to equation 3.

It should be noticed that each synapse may have a different synaptic time constant τ_s and a different time delay Δ, which would affect the shape of its function $\epsilon(s)$, in addition of course to a different weight w. Similarly, each neuron may have a different membrane time constant τ_m and a different threshold θ, which would affect the contribution of its incoming synapses and its own spiking time.

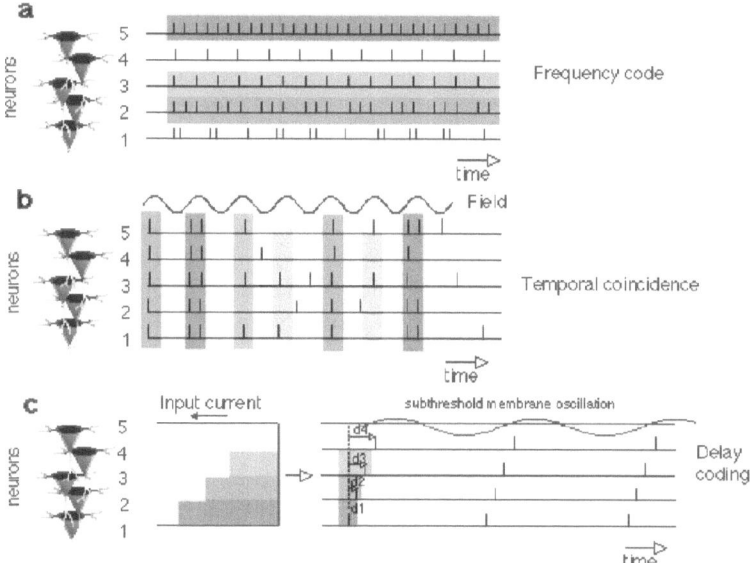

Fig. 4. Some models for encoding sensory information in spiking neurons (from [17]). Fictive spike trains recorded from five imaginary neurons. The different stimulus intensities (represented by gray scale) are converted to different spike sequences. (a) In the frequency code hypothesis [20] neurons generate different frequency of spike trains as a response to different stimulus intensities. (b) In the temporal coincidence hypothesis [22] spike occurrences are modulated by local field oscillation (gamma). Tighter coincidence of spikes recorded from different neurons represent higher stimulus intensity. (c) In the delay coding hypothesis [7] the input current is converted to the spike delay. Neuron 1 which was stimulated stronger reached the threshold earlier and initiated a spike sooner than neurons stimulated less. Different delays of the spikes (d2-d4) represent relative intensities of the different stimulus.

3 Interfacing Spiking Neurons with a Robot

Interfacing a neural network of sigmoid neurons to a robot is relatively straightforward. At regular intervals (100 ms, e.g.), the values read from the sensors of the robot to set the activation values of the corresponding input units. Similarly, the activation values of the output units are read at regular intervals to set the control parameters of the robot actuators, such as the speeds of the wheels.

In a spiking neural network, a single spike is a binary event that can encode only the *presence* or absence of a stimulus. Figure 4 shows three ways to map the *intensity* of sensory information into spiking neurons. A classic method (a) consists of mapping the stimulus intensity to the firing rate of the neuron. This method is based on the hypothesis that a neuron increases its firing rate to indicate stronger stimulation. For example, an often-cited result [2] shows that the firing rate of a stretch receptor in the frog is a monotonically increasing function of the strength of the stimulation and saturates near the maximum firing rate of the neuron.[1] Another method (b) consists of encoding the sensory stimulation across several neurons and mapping the intensity of the stimulation into the number of neurons that spike at the same time. This method is based on the hypothesis that the brain represents meaningful information by synchronizing spiking activities across several neurons [21] and has been supported by measurements in the visual and temporal cortex of monkeys [1,22]. A recently suggested method (c) consists of encoding the strength of the stimulation in the firing delay of the neuron. The underlying hypothesis is that neurons that receive stronger stimulation fire earlier than neurons receiving weaker stimulation and has been supported by measurements in the olfactory neurons [7].

In the experiments reported in this paper, we have used a stochastic version of the firing rate method. In other words, the intensity of the stimulation is represented by the probability that the neuron emits a spike in a time interval. When repetitively measured over relatively long periods of time with respect to the time interval for the same stimulation intensity, the observed firing rate is proportional to the strength of the stimulation.

The transformation of spikes into motor commands presents similar issues. A simple model assumes that muscle stretching is a monotonically increasing function of the firing rate of one or more motor neurons over a short time window (from 20 to 60 ms) [11]. In these experiments, we have taken the firing rate of the motor neurons measured over 20 ms as speed commands to the wheels of the robot.

Another issue is the synchronization between the temporal dynamics of the spiking network and those of the robotic hardware. Biological neurons operate on the millisecond time scale and electronic spiking neurons can operate even faster, but current robotics technology may operate on a slower time scale. For example, in a Khepera robot equipped with a linear camera the visual information can be accessed every 25 ms at optimal lighting conditions, but in order to allow for lower illumination it is safer to access it every 50 or 100 ms. Furthermore, since in our experiments the network of spiking neurons runs on the workstation, we must allow for the time taken by the sensory signals to be transmitted through the serial connection and the time taken to send the motor commands to the wheels of the robot.

One possible solution is to set the elementary unit of time of the spiking neurons to the maximum time interval necessary to read the sensors and set the speeds of the wheels. Another solution is to let the neurons operate at their "natural" time scale (for example, with an update rate of 1 ms) while accessing the robotic interface at regular longer intervals. In this case, the neurons can

---

[1] More detailed studies indicate that this function is best described by a power function of the general form $R = KS^n + C$ where R is the observed firing rate, K is a constant of proportionality, and C is a constant given by the spontaneous firing rate of a neuron in the absence of stimulation (in our experiments, the spontaneous activation is 0) [16].

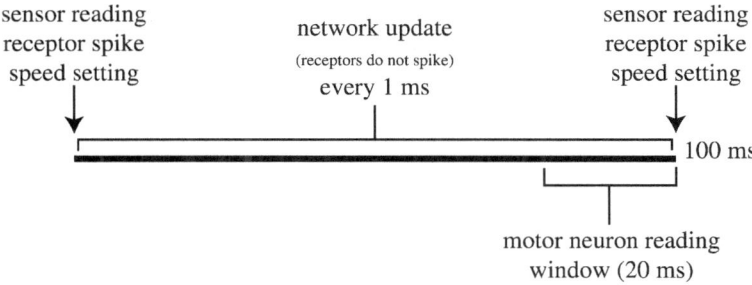

Fig. 5. Time management diagram (see text).

change their states faster than the sensors and actuators of the robot and may be left free to develop internal temporal dynamics.

In these experiments, we have chosen the second option (figure 5). Every 100 ms the sensors of the robot are pre-processed, scaled in the range $[0, 1]$, and used to set the probability of emitting a single spike at that precise time step, while the wheel speeds are set at the end of the 100 ms interval using the firing rate of the motor neurons measured during the previous 20 ms. During the 100 ms interval until the next sensory-motor access, all the neurons in the network are updated every 1 ms (except for the sensory neurons). In the meanwhile the robot moves using constant speed values (the real speed is given by a PID controller). We have included time functions in the code that make sure that the neural network is updated exactly once every millisecond. These functions allow us to keep the system synchronized when other processes are running in the background, when we activate new routines during analysis, and when we use different workstations.[2]

4   Evolution of Vision-Based Navigation

The experiments described in this paper are a preliminary exploration into the evolution of spiking controllers and a comparison with evolution of connection-ist sigmoidal neurons using the same genetic representation and experimental settings.

We have attempted to evolve vision-based controllers for a navigation task in a rectangular arena with textured walls (figure 6). The walls are filled with black and white vertical stripes. The width and spacing of the stripes are random (within the interval $[0.5, 5]$ cm). Given the visual angle of the robot camera and the size of the arena, some areas display higher stripe frequency than other areas.

---

[2] If we had not such functions, a Pentium III at 700 MHz running Linux would update the networks used in our experiments between 1000 and 1200 times per 100 ms, depending on the number of active background processes.

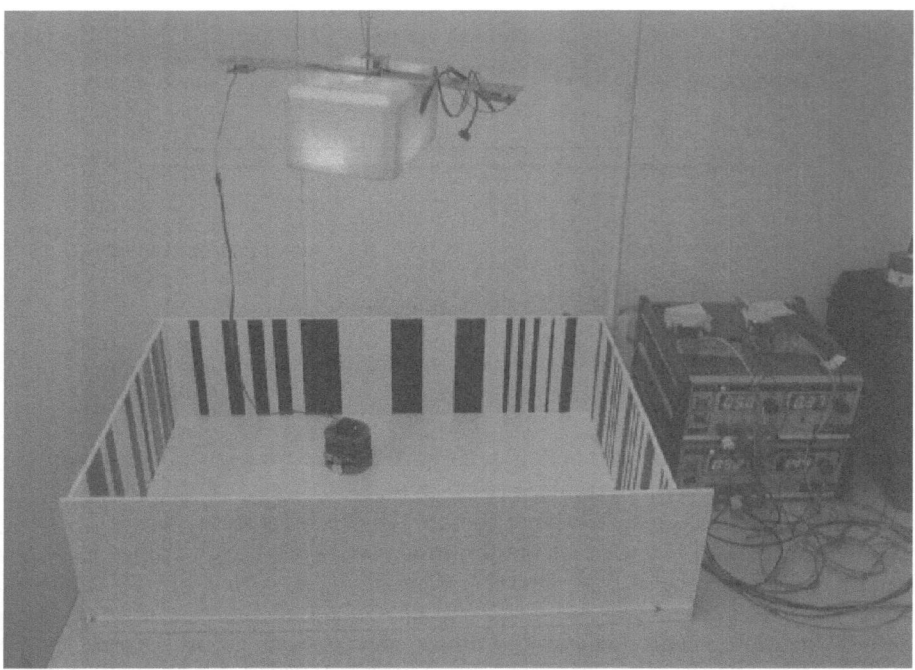

Fig. 6. A Khepera robot equipped with a linear camera is positioned in an arena with black and white vertical stripes of random size painted on the walls at irregular intervals. The arena is lit from above in order to let the evolutionary experiments continue at night. The robot is connected to a workstation through rotating contacts that provide serial data transmission and power supply. The spiking networks and genetic operators run on the workstation. The robot communicates with the workstation every 100 ms.

We have used a Khepera robot equipped with a linear camera module (figure 7). The vision system is composed of a linear array of 64 photoreceptors (left hole) spanning a visual angle of 36 deg and of a light sensor (right hole) used to adjust the sensitivity of the receptors to the global illumination level. Each photoreceptor returns a value between 0 (black) and 255 (white). Given the spacing of the stripes on the wall, we read only 16 photoreceptors equally spaced on the array. These values are convolved with a Laplace filter spanning three adjacent photoreceptors in order to detect contrast. Finally, the convolved image is rectified by taking the absolute values and scaling them in the range $[0, 1]$. The resulting 16 values represent the probabilities of emitting a single spike for each corresponding neural receptor at that precise instant only. The camera values are read every 100 ms (see also figure 5).

In these experiments we use a network of predefined size where the sign of the synaptic weight is given by the sign (excitatory or inhibitory) of the presynaptic neuron. In other words, all the connection coming from an excitatory

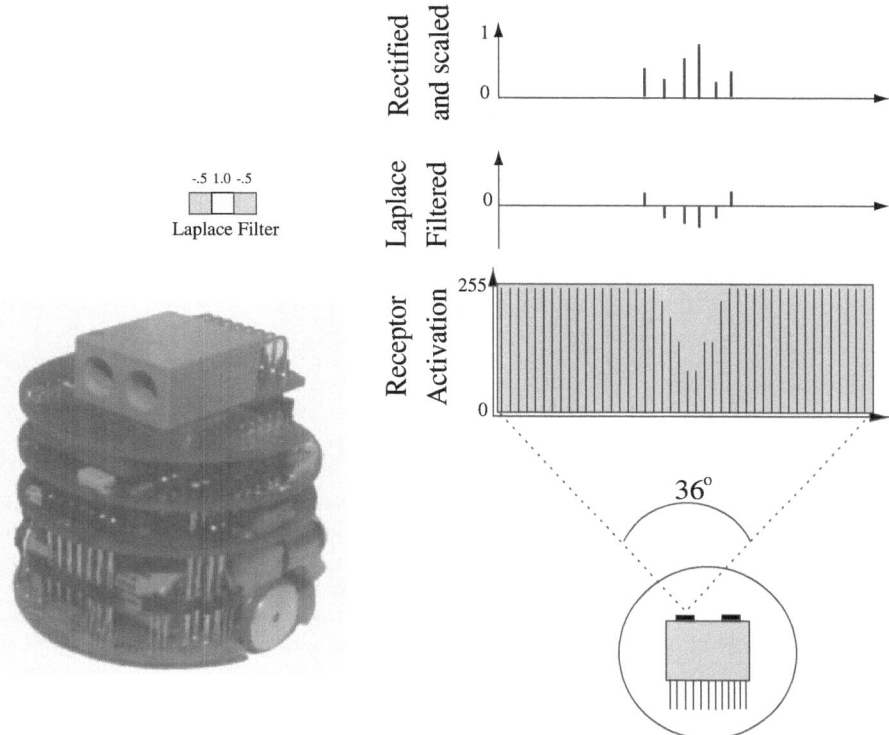

Fig. 7. The Khepera robot equipped with a linear vision system composed of 64 photoreceptors. Only 16 photoreceptors are read every 100 ms and filtered through a Laplace filter in order to detect areas of contrast. The filtered values are transformed into positive values and scaled in the range [0, 1]. These values represent the probability of emitting a spike for each corresponding neural receptor.

(inhibitory) neuron are positive (negative). Sensory receptors are always excitatory. We genetically encode and evolve only the signs of the neurons and the pattern of connectivity among neurons and between neurons and sensory receptors. The network consists of 10 fully-connected neurons, each connected to all sensory receptors (figure 8). There are 18 sensory receptors: 16 transmit vision signals and 2 transmit the error between the desired speeds of the wheels and the speed measured using on-board optical encoders.[3]

Four neurons are used to set the speeds of the wheels in push-pull mode. Two neurons are assigned to each wheel, one neuron setting the amount of forward speed and the other setting the amount of backward speed. The actual speed of the wheel is the algebraic sum of the two speeds. This value is mapped into a

---

[3] The error is transformed in positive values in the range [0, 1] and used a s a probability to emit a spike.

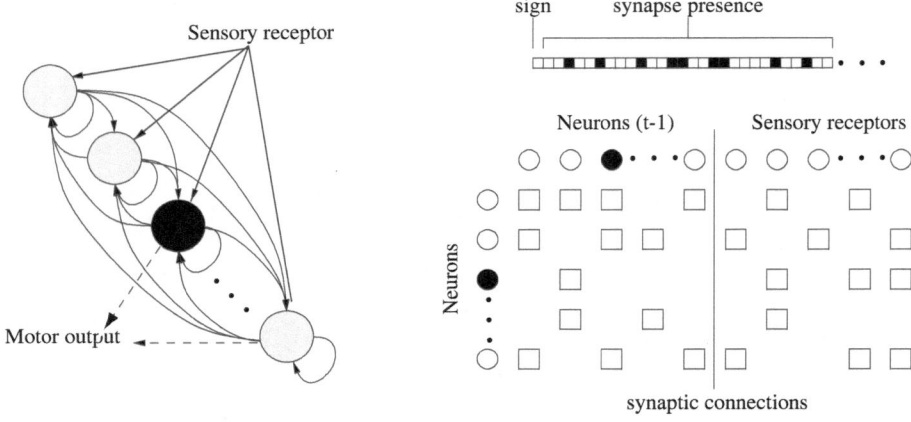

Fig. 8. Architecture of the neural network used in the experiments (only a few neurons and connections are shown) and genetic representation of one neuron. The network has 10 neurons that can be fully connected among each other. In addition, each neuron is connected to 18 spiking receptors (16 for vision and 2 for the error between speed commands and actual speed of the two wheels). Four neurons are used to set the speeds of the two wheels in push-pull mode. *Left*: A conventional representation showing the network architecture. *Right*: The same network unfolded in time (neurons as circles, synaptic connections as squares). The neurons on the column receive signals from connected neurons and receptors shown on the top row. The first part of the row includes the same neurons at the previous time step to show the connections among neurons. Sensory receptors do not have interconnections. The signs of the neurons (white = excitatory, black = inhibitory) and their connectivity is encoded in the genetic string and evolved.

maximum speed of 80 mm/s. However, since the neurons can fire at maximum once every 2 ms (because of the very low negative value taken by the neuron after emitting a spike), in practice the maximum speed is 40 mm/s.

A binary genetic string encodes only the sign of each neuron and the presence of a synaptic connection. The string is composed of n blocks, one for each of the n neurons in the network. The first bit of the block encodes the sign of the neuron and the remaining bits encode the presence/absence of a connection from the n neurons and from the s receptors. Therefore, the total length l of the string is $l = n(1 + n + s)$. In our experiments $l = 10(1 + 10 + 18) = 290$. The other parameters of the neurons and synapses (see section 2) are set as follows: $\theta = 0.1$, $\tau_m = 4$, $\tau_s = 10$, $\Delta = 2$, and $w = 1$ for all neurons and all synapses in the network. The shape of the synaptic and refractory functions for these parameters are shown in figures 1 and 2, respectively. Given these values, only the most recent 20 spikes arriving at each synapse are taken into account for computing the total synaptic contribution according to equation 1. In addition, some noise is added to the refractory period by multiplying the value returned by the refractory function (equation 3) at each time step by a uniformly random value in the range [0, 1]. Preliminary experiments showed that without this added noise, the networks go very quickly into locked oscillations for a very large number of connectivity patterns and parameters.

Each individual of the population is decoded and tested on the robot *two times* for 40 seconds each (400 sensory-motor steps). The fitness function Φ is the sum of the speeds of the two wheels v_{left} and v_{right} measured at every time step t (100 ms) only if both wheels rotate in the forward direction averaged over all T time steps available ($T = 800$)

$$\Phi = \frac{1}{T} \sum_{t}^{T} (v_{left}^t + v_{right}^t) \tag{5}$$

If v_{left} or v_{right} are less than 0 (backward rotation), $\Phi^t = 0$. This fitness function selects individuals for the ability to go as straight as possible while avoiding the walls because it takes a few seconds to travel across the arena and if the robot is stuck against a wall, the wheels can hardly rotate due to the friction on the floor. The fitness function does not use the active infrared sensors available on the robot to judge the distance from the walls because the response profile of these sensors varies depending on the reflection properties of the walls (black stripes reflect approximately 40% less infrared light than white stripes) and on the spectrum component of ambient illumination.

A population of 60 individuals is evolved using rank-based truncated selection, one-point crossover, bit mutation, and elitism [18]. After ranking the individuals according to their measured fitness values, the top 15 individuals produce 4 copies each to create a new population of the same size and are randomly paired for crossover. One-point crossover is applied to each pair with probability 0.1 and each individual is then mutated by switching the value of a bit with probability 0.05 per bit (that is, on average 14.5 bits are mutated for each individual). Finally, a randomly selected individual is substituted by the original copy of the best individual of the previous generation (elitism).

We have run two sets of experiments, each generation taking 80 minutes on the physical robot. Each set of experiments consists of several evolutionary runs starting from a different random initialization of the genetic string. In the first set, we have evolved spiking controllers using the parameters described in this section. The graph on the left of figure 9 shows the average fitness values measured across six runs for 30 generations of evolutionary spiking networks. Since the initial populations are randomly created, the networks display on average 50% random connectivity. This value does not change significantly along generations. The best and average fitness values gradually increase and reach a plateau around the 30th generation.

In a second set of experiments, we have evolved connectionist sigmoid networks. In this case the activation of a neuron $v(i)$ is a value between 0 and 1 computed using the sigmoid function $v(i) = 1/1 + e^{(-A)}$, where $A = \sum_j w_{ij} v_j$. The receptors take on the values that were used as probabilities of emitting a spike for the spiking controllers. The network is updated only one time every 100

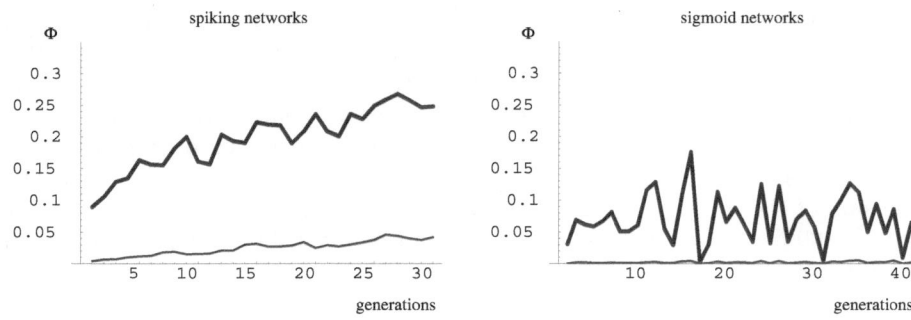

Fig. 9. Fitness values obtained on the physical robot Khepera (best fitness = thick line; average fitness = thin line). Each data point is the average of several evolutionary runs with different random initializations. *Left*: Evolution of spiking networks (average over six runs). *Right*: Evolution of connectionist sigmoid networks where 10 additional generations per run were allowed to check for signs of improvement (average over three runs).

ms.[4] All the other parameters and genetic encoding were identical to those used for the spiking controllers (except for the fact that there is no threshold, synaptic function, and refractory function). The graph on the right of figure 9 shows the average fitness values measured across three runs of connectionist sigmoid networks. Despite allowing for an extra ten generations, none of the evolutionary runs could improve the fitness values along generations. The occasional higher values are given by individuals that perform wide circles independently of the sensory input until they meet a wall where they remain stuck.

The left side of figure 10 shows the architecture and the path of the best spiking controller of the 30th generation during 40 seconds. The right side of the figure shows the corresponding neural activity sampled every 100 ms. The robot moves along a looping trajectory, whose curvature depends on the visual input, without ever remaining stuck against a wall. The behavior does not change when the ambient illumination is increased or decreased since the visual input receives information about contrast, not about the grey levels returned by the photoreceptors.

If the robot is positioned against a wall (at any location), it slowly starts to rotate to the right until it gets away from it. Similarly, if a piece of white paper or one hand is positioned closed enough to its facing direction (2 to 3 cm), it rotates on the spot until it can get away.

---

[4] We have also performed experiments where the sigmoid networks are updated 100 times between two sensory-motor intervals. These experiments generated the same results observed with a single update.

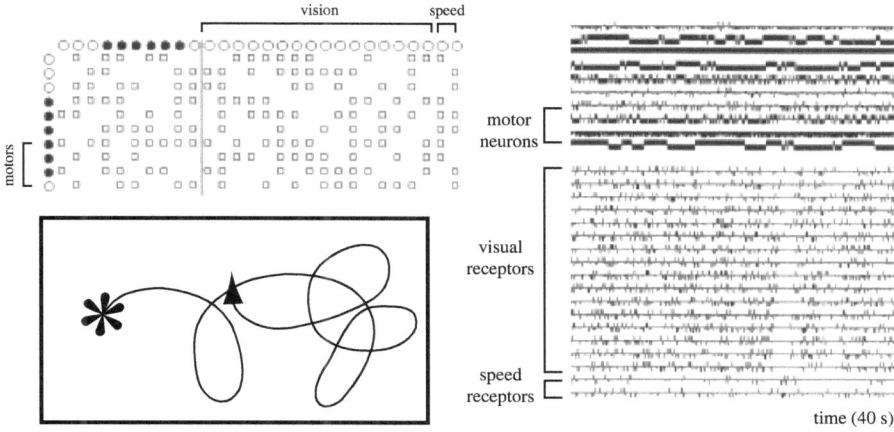

Fig. 10. *Top Left*: Architecture of the best spiking controller after 30 generations. Black circles = inhibitory neurons, white circles = excitatory neurons. *Bottom Left*: Typical trajectory of displayed by this spiking controller. The asterisk indicates the starting point. The curvature of the trajectory depends on the pattern of stripes seen by the robot. *Right*: Corresponding neural activity shown every 100 ms. Each line represents the resting potential of a neuron membrane. Dashes above the line indicate a spike, whereas signs below the line indicate that the neuron is inhibited.

5   Analysis of an Evolved Spiking Controller

In this section we analyze the best spiking controller evolved after 30 generations that has been described above. Since the neural receptors are driven solely by sensory information, do not have interconnections, and can emit a spike only every 100 ms (all spikes are shown on the right plot of figure 10), we focus our analysis on the remaining ten neurons, whose dynamics are updated every 1 ms.

While the robot was freely moving in the environment for 40 seconds, we recorded all the spikes emitted by each neuron. Table 1 shows that several neurons fire at almost maximum rate, which is 500 spikes per second, that is a spike every second millisecond because of the effects of the refractory period. Therefore, these neurons display a consistent self-sustained activity in the Inter Stimulus Interval (100 ms).

Table 1. Average number of spikes per second of the best spiking controller shown in figure 10 measured during 40s of autonomous navigation.

	neuron number									
	1	2	3	4	5	6	7	8	9	10
spikes/s	9	445	453	450	330	40	129	363	0	452

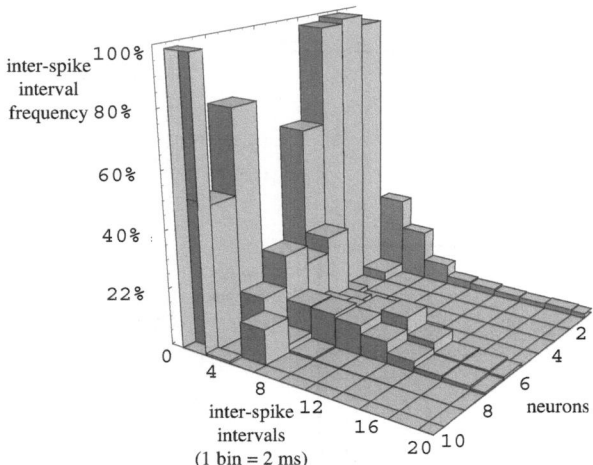

Fig. 11. Inter-spike interval frequencies measured during 10 seconds. The time intervals between adjacent spikes of a neuron are counted, grouped in incremental bins of 2 ms (2, 4, 6, 8, ..., 20), and normalized by the total number of spikes emitted by the neuron.

The firing rates of neurons 7–10 measured over 20 ms are used to set the speeds of the wheels. The firing rate of neurons 7 and 8 respectively set the backward and forward speed components of the right wheel, whereas the firing rates of neurons 9 and 10 set the backward and forward speed components of the left wheel. The spiking rates of these motor neurons suggests that the neural network controls the turning angle of the robot by changing the rotation speed of the right wheel (controlled by neurons 7 and 8) while the left wheel is kept at constant forward speed (by neuron 10).

The relative frequency of time intervals between spikes of each neuron provides further information on the the internal dynamics of the network (figure 11). This indicator is obtained by measuring the lengths of the intervals between two adjacent spikes over a 10 seconds period and grouping the occurrences of these intervals in bins of 2 ms (2 ms, 4 ms, 6 ms, 8 ms, etc.). These values are then divided by the total number of spikes fired by each neuron over 10 seconds in order to obtain a relative frequency. For example, a value of 0.2 in the third interval bin, means that that 20% of the spikes emitted by the neuron occur at intervals of 6ms. As expected from the firing rates shown in table 1, the data of figure 11 show that most neurons always fire at short intervals (2 and 4 ms). For example, this is the case of neuron 10 whose regular firing every 2 ms sustains constant forward speed of the left wheel. Some neurons fire also at longer intervals, but these intervals are rarely longer than 10 ms. Therefore, the dynamics of the evolved network settle into a stable state within approximately 10 ms after receiving sensory stimulation. To check this hypothesis, we have slowed down

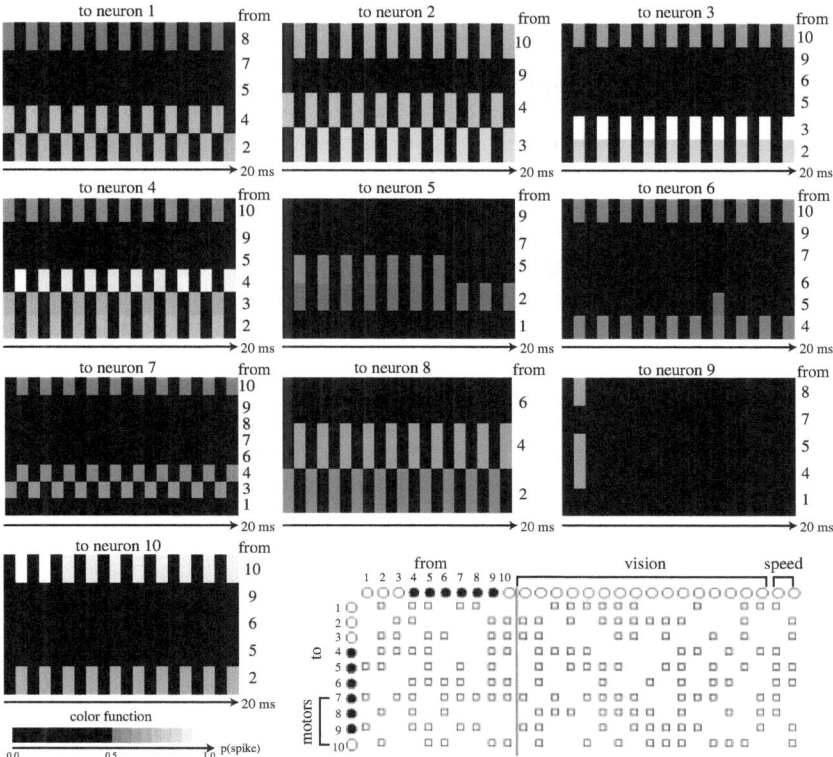

Fig. 12. Temporal Spike Correlographs. For each spike emitted by a neuron during a 10 s window, we count the occurrences of spikes emitted at within the preceding 20 ms (+2 ms to account for synaptic delay) by its pre-synaptic neurons, group them in 20 bins of 1 ms each, and normalize them by the total number of spikes emitted by the post-synaptic neuron.

the number of network updates between sensory updates from 100 to 50 and did not observe any difference in the behavior of the robot.

Since each neuron receives, on average, signals from five other neurons (without counting the connections from the sensory receptors), it is worth asking whether all the pre-synaptic neurons play the same role in causing a spike. To answer this question, we have developed the Temporal Spike Correlographs (figure 12). The correlations between the firing time of a neuron and the firing times of its pre-synaptic neurons within the 20 ms window preceding a spike[5] provide an indication of the most important synaptic channels. These measures, shown in figure 12, are obtained as follows for each neuron. When the neuron emits

---

[5] Remember that this is the time window of synaptic integration used in these experiments; see also figure 1.

a spike, we record the presence of a spike coming from each of its pre-synaptic neurons within a time window of 20 ms displaced by 2 ms in order to account for the synaptic delay (see equation 1 and figure 1). We repeat these measures over a period of 10 seconds arbitrarily chosen while the robot moves in the environment and count how many times at each millisecond there has been a spike from each connected pre-synaptic neuron. These counts are then divided by the total number of spikes emitted by the post-synaptic neuron. We take these values as the probability that pre-synaptic neuron j emits a spike t ms earlier than post-synaptic neuron i. If a neuron is connected to, say, 5 presynaptic neurons, we will have 5 series of 20 probabilities each. Figure 12 shows these probabilities as gray levels where white represents the highest probability and black represents probabilities lower or equal than 0.5.

These probability distributions tell us a number of things. To start with, not all synaptic channels are equally important. For example, when neuron 3 fires there is a probability lower or equal than 0.5 that neurons 5, 6, and 9 at any instant within the 20 ms window of synaptic sensitivity. On the other hand, we are almost certain of finding a spike every second millisecond coming from neurons 2, 3 (self-connection), and, to a lesser extent, from neuron 10. Considering that the threshold of the neurons is set to 0.1, a single spike emitted 6 ms earlier (including the time delay not shown in figure 12) by an excitatory neuron could cause a post-synaptic spike. In practice, things are more complex because we should also consider the interplay of excitation and inhibition. However, to a first approximation we can speculate that synaptic channels shown in black do not play an important role in controlling the behavior of the neuron.

If we weight the pattern of connectivity among neurons by the importance of the connections, we get an idea of which neurons are not playing an important role in the network. Neurons 1, 5, 6, and 9[6] either do not make synaptic connections *to* other neurons or, when they do, their contribution does not play an important role in the firing of other neurons (as indicated by the black rows corresponding to these neurons). In the next section, we will explore these hypotheses by performing lesion studies.

Among the remaining six neurons, three (3, 4, and 10) have self-connections with high spike correlation. Notice that a self-connection does not necessarily imply high spiking correlation because of the effects of inhibitory channels and/or of long inter spike intervals (this is the case of neuron 5 and 6, for example). In the case of neurons 3 and 10 these self-connections are sufficient to generate a regular spiking activity independently of other inputs because there are not "important" inhibitory connections and the membrane threshold of 0.1 can be easily exceeded by their self-generated train of spikes over the 20 ms window (see also figure 1). Instead, the firing rates of neurons 2, 7, and 8 are determined by a delicate balance of inhibition and excitation coming from other neurons (and from neural receptors).

Another question is whether the evolved controller exploits spatio-temporal spike correlations from pre-synaptic neurons, as those postulated by the theories

---

[6] Neuron 9 spiked only 3 times during the 10 seconds of observation.

of cortical activity based on coincidence detection [1,22,12]. If this is the case, for a given neuron we should observe a precise pattern of spikes across different pre-synaptic neurons that occurs regularly. From the distribution of inter spike intervals (figure 11), we know that this pattern must occur within the 15 ms window preceding the spike of the neuron. Although the plots of figure 12 clearly show regular occurrences of pre-synaptic spikes, there is not enough evidence that post-synaptic firing is caused by the coincidence of these spikes across different neurons. We can certainly rule it out for neurons 3 and 10 whose activity is mainly self-generated. We can also almost certainly rule it out for other neurons that fire mostly every second millisecond (2 and 4) because this time is insufficient to detect spatio-temporal patterns. Finally, we can rule it out for neuron 9 that almost never spikes. For the remaining five neurons, the analysis performed above is not sufficient to tell whether these patterns exist and play a role because their firing condition is the result of complex interactions between inhibitory and excitatory spikes. The tests described in the next subsection will help us to answer this question.

5.1   Lesion Studies

In the previous section we have speculated that neurons 1, 5, and 6 do not play a major role in the network on the basis of their "low contribution" to the firing of other neurons in the network. To check that hypothesis, we have systematically lesioned one neuron at a time and tested the lesioned controller in the environment three times for a duration of 80 seconds each. A lesion is done by silencing completely the output of the neuron. When neurons 1, 5, *or* 6 were lesioned the robot reported an average fitness of 0.22, practically identical to the fitness measured before the lesion (0.23), and no behavioral change was observed. We then lesioned all neurons 1, 5, *and* 6 and performed the same test. In this case the average reported fitness was 0.19, only slightly less than that observed before the multiple lesion and the behavior was almost the same. However, this time the trajectory was more straight when far from the walls and more jerky near the walls. This is probably due to the fact that the combined spiking rates of these neurons provide non-specific signal boost to the internal activity of the network. When they are lesioned all together, the robot behavior is slightly more dependent on the signals sent by visual receptors than on the internal activity.

Instead, when neuron 2 is lesioned, the robot rotates on the spot without influence of the visual input. Conversely, when neuron 3 is lesioned, the robot moves on a more or less straight trajectory (depending on the visual pattern) until the nearest wall. In both cases, the average fitness in the three tests is close to 0.0. Borrowing the method of double dissociation from cognitive neuropsychology [4], we can conclude that neuron 2 is responsible for moving straight on the basis of visual information, whereas neuron 3 generates a turn when the robot is gets too close to the walls. The role of neuron 2 is further supported by the pattern of connectivity from the visual receptors. It is the only neuron that has a full connection from the six vision receptors exactly in the middle

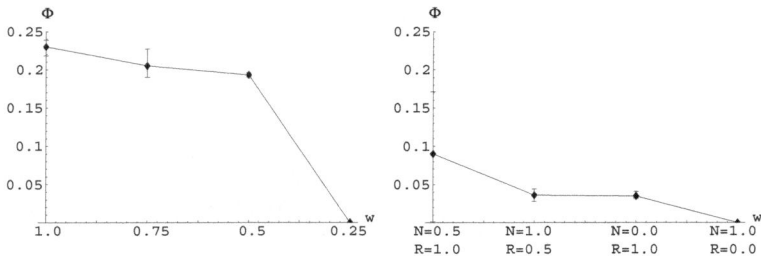

Fig. 13. Performance tests with synaptic decay. Data points represent average values (error bars indicate lowest and highest values) over three runs of 80 seconds each starting from a different location. The fitness value below which the robot is no longer capable of avoiding walls is approximately 0.15. *Left*: Synaptic strength is uniformly set to lower values w for all connections in the network. The first data point represents the performance of the evolved controller with intact synaptic strength ($w = 1$). *Right*: Synaptic strength is separately decreased for the synapses coming from the neurons (N) and for the synapses coming from the receptors (R).

of the visual field. Finally, when neuron 4 is lesioned the robot displays almost the same behavior, but cannot avoid the walls if it gets closer than 3 or 4 cm to them. Since neuron 4 is inhibitory and affects significantly both neuron 2 and motor neurons 7 and 8 (that control the right wheel to steer the robot), in normal conditions its role may be that of detecting a near collision and send stronger inhibition to neuron 2, which would result in a sharper turn.

5.2   Synaptic Decay

Another way to lesion the network consists of changing the strengths of the synaptic connections. These tests are of particular interest for hardware implementation of spiking neurons because the storage of weights for adaptive synapses is a major problem. Synaptic strengths are often stored and adapted by means of coupled capacitors whose voltages tend to decay relatively quickly. The same problem is found in analog VLSI implementations of adaptive weights. For example, Lewis et al. [13] report that their aVLSI neuromorphic Central Pattern Generator can store the values of synaptic weights for a few seconds with a decay rate of 0.1 V/s.

In order to assess the robustness of the evolved spiking controller with decaying synaptic strengths, we have performed three tests with weaker synaptic weights, at 0.75, 0.5, and 0.25 respectively. Each test consists of setting all the synaptic strengths (from the neurons and from the receptors) to a fixed value and measuring the fitness of the robot across three trials of 80 ms each starting at a different location in the environment. The average fitness values are plotted in the left graph of figure 13. The first data point shows the fitness of the evolved controller with its original synaptic strengths ($w = 1.0$). To some

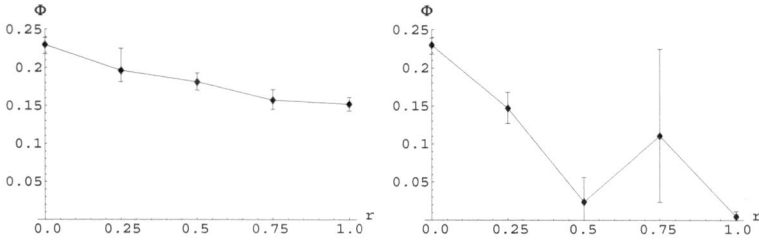

Fig. 14. Performance tests with noisy synaptic decay. Data points represent average values (error bars indicate lowest and highest values) over three runs of 80 seconds each starting from a different location. The fitness value below which the robot is no longer capable of avoiding walls is approximately 0.15. *Left*: A random value from the interval 0, r is subtracted from the original synaptic strength ($w = 1$) every millisecond. The random values are generated anew every millisecond and are different for every synapse. *Right*: Same as for the left graph, but the random values are subtracted only at the beginning of the test and the resulting weights are maintained constant for the whole duration of the test.

approximation, we can say that fitness values above 1.5 correspond to behaviors that retain the wall avoidance strategies and, to some extent, the trajectories of the original evolved controller. These tests indicate that the spiking controller can withstand uniform weight decay up to 50% of its original values. The lower fitness values are caused by slower movement or slightly jerkier movements.

In another series of tests, we have decreased the strengths separately for the synaptic connections among the neurons (N) and for the synaptic connections from the receptors (R). Four tests have been performed with the following combinations of strengths: $N = 0.5, R = 1.0$, $N = 1.0, R = 0.5$, $N = 0.0, R = 1.0$, and $N = 1.0, R = 0.0$. In none of these conditions the spiking controller could manage to navigate around the environment (graph on the left of figure 13). When only the connections among the neurons are lowered to 0.5 the robot goes straight but cannot avoid walls and when they are set to 0 it turns on itself. When only the connections from the sensors are lowered to 0.5 the robot turns on itself (but in the opposite direction than that used during normal behavior) and, of course, when they are set to 0.0 it does not move. These tests, combined with those on uniform synaptic decay, indicate that the evolved controller can withstand severe decay only if it is uniformly applied to all synapses in the network.

In another series of tests, we have studied the effects of noisy synaptic variations that may be caused, for example, by external radiation in analog VLSI microchips. In the first set of tests we change the strengths of each synapses every millisecond by subtracting a uniformly random number taken from the interval $[0, r]$ from the original strength ($w = 1$). The random numbers are generated separately for each synapse at each millisecond. Four tests have been performed with $r = 0.25$, $r = 0.5$, $r = 0.75$, and $r = 1.0$, respectively. Each test consists of three experiments of 80 seconds each, as above. The average fitness data shown on the left graph of figure 14 (the first data point corresponds to a control condition without noise) indicate that the performance of the evolved controller degrades very slowly, but always preserves the same navigation abilities and overall trajectories. The small degradation is caused by the appearance of slightly jerky movements for $r = 0.25$ and lower speed for $r = 0.25$ and $r = 0.75$. For $r = 1.0$, we begin to observe straighter trajectories while still preserving the ability to avoid walls.

These results definitely rule out the possibility that the evolved controller may exploit spatio-temporal spike correlations because, if this were the case, the severe noise conditions of these tests would disrupt the precise contributions of individual spikes across the synaptic channels and would result in bad performance.

Since the synaptic noise is applied anew every millisecond, the average value over relatively long intervals approximates $r/2$. Therefore, if the network dynamics are based on firing rate, instead of firing time, the effects of this type of noise should be equivalent to those obtained by subtracting from each synapse a fixed value equal to $r/2$. This is indeed what we observe when we compare the results of these tests with those obtained with uniform synaptic decay (graph on the left of figure 13).

In a second set of tests with noisy synapse, we subtract from each synaptic strength a uniformly random number taken from an interval $[0, r]$ at the beginning of the test and maintain the resulting values for the whole duration of the test. Using the same procedure described above, we perform four tests with $r = 0.25$, $r = 0.5$, $r = 0.75$, and $r = 1.0$, respectively. The results shown in the graph on the right of figure 14 indicate that the controller is badly affected by non-uniform synaptic decrements. With the exception of $r = 0.25$, where the robot is still capable of maintain the basic navigation abilities, all the remaining conditions produce robots that mostly rotate on place. This last set of data fits the observations obtained with non-uniform synaptic decay across synaptic groups (neurons and receptors).

To summarize the results on synaptic lesions, we can conclude that the evolved controller can withstand strong synaptic decay, even if this is generated by a random process, as long as the decay is uniformly applied across the whole network.

6   Discussion

Despite the complex dynamics of spiking neurons and the higher number of free parameters with respect to sigmoid neurons, the results reported in this paper indicate that evolution can easily discover functional networks by searching the space of connectivity. Although the values of the neuron parameters have been set according to data reported in the literature [6] without attempting to optimize them for this specific implementation, we cannot exclude that different values would make the network harder or easier to evolve. As we already mentioned, the presence of noise in the refractory period is the only factor that we found crucial for the evolvability of functional networks of spiking neurons. Since this type of noise affects the recovery time of the neuron, it reduces the probability that the networks fall into a state of locked oscillations that cannot be disrupted by sensory inputs.

The experimental data also show that it is harder to find functional networks of sigmoid neurons evolved under the same constraints. This does not mean that it is impossible to evolve networks of sigmoid neurons for this task, but that such networks probably require the variation and precise definition of other parameters, such as the strengths of the connections (in these experiments, they were all set to 1). It may also be the case that networks of sigmoid neurons with synaptic time delays and continuous dynamics [3] could be evolved under the same conditions used for spiking neurons. Even if that was the case, we think that spiking neurons represent a more powerful substrate for real-time mapping of sensory information into motor actions because these networks can potentially function both firing rate and firing time modes. In firing time mode, the time of arrival of a spike (or of a precise pattern of spikes distributed across several synapses) would be sufficient to trigger an appropriate response all the way up to the motor commands allowing the robot to respond in a few milliseconds to complex sensory patterns.

We should then ask why we did not find evidence that the neurons of our evolved spiking controllers responded in firing time mode (although we cannot exclude that some of their ancestors did). There could be several reasons for the evolutionary choice of firing rate instead of firing time. One of them is the relatively low membrane threshold $\theta = 0.1$ used in these experiments. Considering the values chosen for the other parameters, here a post-synaptic spike can be triggered by a single spike emitted by an excitatory pre-synaptic neuron 5 ms earlier (including the synaptic time delay). Therefore, it is unlikely that the neurons require multiple spikes across different synaptic channels in order to become active. A solution could be to either set higher threshold values, use smaller synaptic weights, or change the shape of the synaptic integrator (equation 1). Of course, these parameters could also be genetically encoded and evolved along with the pattern of connectivity.

Another reason is that the experimental settings are such to bias the system towards firing rate mode. Here the network is allowed 100 iterations between sensory-motor commands, which is sufficient to send several spikes along a connection and thus encode information in firing rate mode. Furthermore, the speeds of the robot wheels are set using the firing rate measured over a window of 20 ms. It would be interesting to evolve spiking controllers for robots where both sensory and motor dynamics match the dynamics of the spiking network. At the sensory level, these features could be provided by neuromorphic vision systems [9] whereas, at the motor level, one could use the spikes to directly drive the motors.

We also think that the way in which the sensory information is encoded into spiking neurons may affect the evolved functioning modality. In section 3 we have discussed some models of sensory encoding for spiking neurons. Whereas sensory information encoded with the frequency model or the temporal coincidence model may be "understood" using a firing rate mode by simply summing up the quantity of incoming spikes across different channels over some time window, the delay coding model may force post-synaptic neurons to pay attention to the time of arrival of spikes in order to disambiguate the nature of the sensory stimulation.

Despite the relatively general form of the model of spiking neurons used in these experiments, one may not need all this level of detail when it comes to evolve networks of spiking neurons to be implemented in physical circuits. In that case, a simpler Integrate-and-Fire model which is relatively easier to realize with analog VLSI technology may provide similar functionalities.

7   Conclusion

The experiments described in this article are a preliminary exploration into the evolvability and properties of spiking neurons as control systems for autonomous robots. We have shown that artificial evolution can quickly discover small networks of spiking neurons to perform a non-trivial vision-based navigation task. The comparison with networks of sigmoid neurons indicate that, all the factors being equal, the intrinsic dynamics of spiking neurons provide more degrees of freedom that can be exploited by evolution to generate viable controllers.

The analysis tools described in this paper have allowed us to understand the roles played by individual neurons in the evolved network and make predictions about the influence of their activity on the behavior independently of their observed firing rate. These predictions have been confirmed with lesion studies on single neurons and groups of neurons.

The evolved network displays remarkable robustness in face of constant or irregular synaptic decay as long as this happens more or less uniformly across the entire network. Since this assumption is quite reasonable in the case of hardware implementation with adaptive or evolvable synaptic weights, we believe that it represents an interesting method for the evolution of analog VLSI spiking circuits characterized by tiny size and extremely small energetic consumption. This technology could be used for micro autonomous robots or for flying robots where payload and energetic autonomy are a major constraint.

Acknowledgements. This work was supported by the Swiss National Science Foundation, grant no. 620–58049.

References

1. M. Abeles. *Corticonics*. Cambridge University Press, Cambridge, 1991.
2. E. D. Adrian. *The basis of sensation: The action of the sense organs*. W. W. Norton, New York, 1928.
3. R. D. Beer and J. C. Gallagher. Evolving dynamical neural networks for adaptive behavior. *Adaptive Behavior*, 1:91–122, 1992.

4. A. W. Ellis and A. W. Young. *Human Cognitive Neuropsychology*. Lawrence Erlbaum Associates, London, 1988.
5. W. Gerstner. Associative memory in a network of biological neurons. In R. P. Lippmann, J. E. Moody, and D. S. Touretzky, editors, *Advances in Neural Information processing Systems 3*, pages 84–90. Morgan Kaufmann, San Mateo, CA, 1991.
6. W. Gerstner, J. L. van Hemmen, and J. D. Cowan. What matters in neuronal locking? *Neural Computation*, 8:1653–1676, 1996.
7. J. J. Hopfield. Pattern recognition computation using action potential timing for stimulus representation. *Nature*, 376:33–36, 1995.
8. G. Indiveri. A Neuromorphic VLSI device for implementing 2D selective attention systems. *IEEE Transactions on Neural Networks*, page in press, 2001.
9. G. Indiveri and R. Douglas. ROBOTIC VISION: Neuromorphic Vision Sensors. *Science*, 288:1189–1190, 2000.
10. G. Indiveri and P. Verschure. Autonomous vehicle guidance using analog VLSI neuromorphic sensors. In W. Gerstner, A. Germond, M. Hasler, and J-D. Nicoud, editors, *Proceedings of the 7th International Conference on Neural Networks*, pages 811–816. Springer Verlag, Berlin, 1997.
11. E. R. Kandel, J. H. Schwartz, and T. M. Jessell. *Principles of Neural Science*. 4th edition. McGraw-Hill Professional Publishing, New York, 2000.
12. P. Koenig, A. K. Engel, and W. Singer. Integrator or coincidence detector? the role of cortical neuron revisited. *Trends in Neuroscience*, 19:130–137, 1996.
13. M. A. Lewis, R. Etienne-Cummings, A. H. Cohen, and M. Hartmann. Toward biomorphic control using custom aVLSI CPG chips. In *Proceedings of IEEE International Conference on Robotics and Automation*. IEEE Press, 2000.
14. W. Maas and C. M. Bishop, editors. *Pulsed Neural Networks*. MIT Press, Cambridge, MA, 1999.
15. C. Mead. *Analog VLSI and Neural Systems*. Addison-Wesley, Reading, MA, 1991.
16. V. B. Mountcastle, G. F. Poggio, and G. Werner. The relation of thalamic cell response to peripheral stimuli varied over an intensive continuum. *Journal of Neurophysiology*, 26:807–834, 1963.
17. Z. Nadasdy. *Spatio-Temporal Patterns in the Extracellular Recording of Hippocampal Pyramidal Cells: From Single Spikes to Spike Sequences*. PhD thesis, Rutgers University, Newark, NJ, 1998. available at http://osiris.rutgers.edu/Buzsaki/Posters/Nadasdy/thesis.html.
18. S. Nolfi and D. Floreano. *Evolutionary Robotics: Biology, Intelligence, and Technology of Self-Organizing Machines*. MIT Press, Cambridge, MA, 2000.
19. D. E. Rumelhart, J. McClelland, and PDP Group. *Parallel Distributed Processing: Explorations in the Microstructure of Cognition: Foundations*. MIT Press-Bradford Books, Cambridge, MA, 1986.
20. C. S. Sherrington, editor. *Integrative Action of the Nervous System*. Yale University Press, New Haven, CT, 1906.
21. W. Singer. Search for coherence: a basic principle of cortical self-organization. *Concepts in Neuroscience*, 1:1–26, 1990.
22. W. Singer and C. M. Gray. Visual feature integration and the temporal correlation hypothesis. *Annual Review of Neuroscience*, 18:555–586, 1995.
23. A. Villa. Empirical evidence about temporal structure in multi-unit recordings. In R. Miller, editor, *Time and the Brain*. Harwood Academic Publishers, Reading, UK, 2000.

First Three Generations of Evolved Robots

Jordan B. Pollack, Hod Lipson, Pablo Funes, and Gregory Hornby

Computer Science Dept.,
Brandeis University,
Waltham,
MA 02454, USA
pollack@cs.brandeis.edu

Abstract. The field of robotics today faces an economic predicament: most problems in the physical world are too difficult for the current state of the art. The difficulties associated with designing, building and controlling robots have led to a stasis, and robots in industry are only applied to simple and highly repetitive manufacturing tasks. Over the last few years we have been trying to address this challenge through an alternative approach: Rather than a seeking an intelligent general-purpose robot, we are seeking the process that can automatically design and fabricate special purpose mechanisms and controllers to achieve specific short-term objectives. This short paper provides a brief review of three generations of our research results. Automatically designed high part-count static structures that are buildable, automatically designed and manufactured dynamic electromechanical systems, and modular robots automatically designed through generative encoding. We expect that with continued improvement in simulation, manufacturing, and transfer, we will achieve the ability to automatically design and fabricate custom machinery for short-term deployment on specific tasks.

1 Roboeconomics

The field of Robotics today faces a practical economic problem: Flexible machines with minds cost so much more than manual machines and their humans operators. Few would spend $2K on a vacuum cleaner when a manual one is $200, or half a million dollars on a driverless car when a regular car is $20K, plus $6\frac{1}{2}$ per hour for its driver. The high costs associated with designing, building and controlling robots have led to a stasis [29] and robots in industry are only applied to simple and highly repetitive manufacturing tasks. Even though sophisticated teleoperated machines with sensors and actuators have found important applications (exploration of inaccessible environments for example), they leave very little decision, if at all, to the on-board software [30]

The central issue addressed by our work is a low cost way to get a higher level of complex physicality under control. We seek more controlled and moving mechanical parts, more sensors, more nonlinear interacting degrees of freedom — without entailing both the huge fixed costs of human design and programming and the variable costs in manufacture and operation. We suggest that this can

T. Gomi (Ed.): ER 2001, LNCS 2217, pp. 62–71, 2001.

be achieved only when robot design and construction are fully automatic, and the results are inexpensive enough to be disposable.

Traditionally, robots are designed on a disciplinary basis: Mechanical engineers design complex articulated bodies, with state-of-the-art sensors, actuators and multiple degrees of freedom. These elaborate machines are then thrown over the wall to the control department, where software programmers and control engineers struggle to design a suitable controller. Even if an intelligent human can learn to control such a device, it does not follow that automatic autonomous control can be had at any price. Humans drastically underestimate animal brains: Looking into nature we see animal brains of very high complexity, brains more complex than the bodies they inhabit, controlling bodies which have been selected by evolution precisely because they were controllable by those brains. In nature, the body and brain of a horse are tightly coupled, the fruit of a long series of small mutual adaptations — like chicken and egg, neither one was designed first. There is never a situation in which the hardware has no software, or where a growth or mutation — beyond the adaptive ability of the brain — survives. The key is thus to evolve both the brain and the body, simultaneously and continuously, from a simple controllable mechanism to one of sufficient complexity for a particular specialized task.

The focus of our research is how to automate the integrated design of bodies and brains using a co-evolutionary learning approach. The key is to evolve both the brain and the body — simultaneously — from a simple controllable mechanism to one of sufficient complexity for a task. Within a decade we see three technologies that are maturing past threshold to make this a possible new industry. One is the increasing fidelity of advanced mechanical design simulation, stimulated by profits from successful software competition [35]. The second is rapid, one-of-a-kind prototyping and manufacture, which is proceeding from 3D plastic layering to stronger composite and metal (sintering) technology [8]. The third is our understanding of coevolutionary machine learning in design and intelligent control of complex systems, which is the primary specialty of our laboratory [1,19,32].

2   Coevolution

Coevolution, when successful, dynamically creates a series of learning environments each slightly more complex than the last, and a series of learners which are tuned to adapt in those environments. Sims' work (1994) on body-brain coevolution and the more recent Framsticks simulator [22] demonstrated that the neural controllers and simulated bodies could be coevolved. The goal of our research in coevolutionary robotics is to replicate and extend results from virtual simulations like these to the reality of computer designed and constructed special-purpose machines that can adapt to real environments. We are working on coevolutionary algorithms to develop control programs operating realistic physical device simulators, both commercial-off-the-shelf and our own custom simulators, where we finish the evolution inside real embodied robots. We are

ultimately interested in mechanical structures that have complex physicality of
more degrees of freedom than anything that has ever been controlled by human designed algorithms, with lower engineering costs than currently possible
because of minimal human design involvement in the product.

It is not feasible that controllers for complete structures could be evolved (in
simulation or otherwise) without first evolving controllers for simpler constructions. Compared to the traditional form of evolutionary robotics [6,9,13,21,27]
which serially downloads controllers into a piece of hardware, it is relatively easy
to explore the space of body constructions in simulation. Realistic simulation is
also crucial for providing a rich and nonlinear universe. However, while simulation creates the ability to explore the space of constructions far faster than real-
world building and evaluation could, there remains the problem of transfer to real
constructions and scaling to the high complexities used for real-world designs.

3   Research Products

We describe three major projects in achieving Fully Automated Design (FAD)
and manufacture of high-parts-count autonomous robots. The first is evolution
inside simulation, but in simulations more and more realistic so the results are
not simply visually believable, as in Sims' work (1994), but also buildable. We
investigated transferring evolved high part-count, static structures from simulation to the real world. The second is to evolve automatically buildable dynamic
machines that are nearly autonomous in both their design and manufacture.
The third thrust, and perhaps hardest, addresses scaling to more complex tasks:
handling complex, high part-count structures through modularity. We have preliminary and promising results in each of these areas, which we outline below.
Further work in automatic manufacturing, sensor integration, and task specification will follow.

3.1   Evolving Buildable Mechanical Simulations (Legobots)

The idea of evolving mechanical structures inside simulation is not new, and
Commercial CAD simulation systems are exciting, yet not constrained enough
for the products to be buildable, because a human provides numerous constraints
to describe reality. In order to evolve both the morphology and behavior of
autonomous mechanical devices that can be built, one must have a simulator
that operates under many constraints, and a resultant controller that is adaptive
enough to cover the gap between the simulated and real world. Features of a
simulator for evolving morphology are:

 - Representation — should cover a universal space of mechanisms.
 - Conservative — because simulation is never perfect, it should preserve a
   margin of safety.
 - Efficient — it should be quicker to test in simulation than through physical
   production and test.
 - Buildable — results should be convertible from a simulation to a real object.

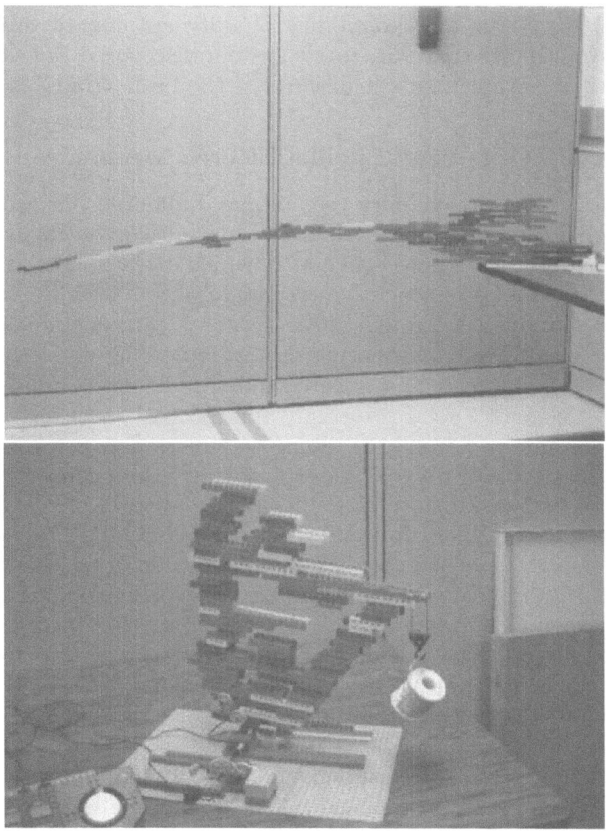

Fig. 1. Evolved Cantilevered Bridge, and Triangular Crane. Only Static components are fully designed by evolutionary algorithms.

One approach is to custom-build a simulator for modular robotic components, and then evolve either centralized or distributed controllers for them. In advance of a modular simulator with dynamics, we recently built a simulator for (static) Lego bricks, and used very simple evolutionary algorithms to create complex Lego structures, which were then manually constructed ([10]-[12]). Our model considers the union between two bricks as a rigid joint between the centers of mass of each one, located at the center of the actual area of contact between them. This joint has a measurable torque capacity. That is, more than a certain amount of force applied at a certain distance from the joint will break the two bricks apart. The fundamental assumption of our model is this idealization of the union of two Lego bricks. The genetic algorithm reliably builds structures that meet simple fitness goals, exploiting physical properties implicit in the simulation. Building the results of the evolutionary simulation (by hand) demonstrated the power and possibility of fully automated design. The long bridge of Figure 1 shows that our simple system discovered the cantilever, while the weight-carrying crane shows it discovered the basic triangular support.

3.2   Genetically Organized Lifelike Electro Mechanics (GOLEM)

The next step is to add dynamics to modular buildable physical components, and to insert their manufacturing constraints into the evolutionary process. We are experimenting with a new process in which both robot morphology and control evolve in simulation and then replicate automatically into reality [24]. The robots are comprised of only linear actuators and sigmoidal control neurons embodied in an arbitrary thermoplastic body. The entire configuration is evolved for a particular task and selected individuals are printed pre-assembled (except motors) using 3D solid printing (rapid prototyping) technology, later to be recycled into different forms. In doing so, we establish for the first time a complete physical evolution cycle. In this project, the evolutionary design approach assumes two main principles: (a) to minimize inductive bias, we must strive to use the lowest level building blocks possible, and (b) we coevolve the body and the control, so that they stimulate and constrain each other. We use arbitrary networks of linear actuators and bars for the morphology, and arbitrary networks of sigmoidal neurons for the control. Evolution is simulated starting with a soup of disconnected elements and continues over hundreds of generations of hundreds of machines, until creatures that are sufficiently proficient at the given task emerge. The simulator used in this research is based on quasi-static motion. The basic principle is that motion is broken down into a series of statically-stable frames solved independently. While quasi-static motion cannot describe high-momentum behavior such as jumping, it can accurately and rapidly simulate low-momentum motion. This kind of motion is sufficiently rich for the purpose of the experiment and, moreover, it is simple to induce in reality since all real-time control issues are eliminated. Several evolution runs were carried out for the task of locomotion. Fitness was awarded to machines according to the absolute average distance traveled over a specified period of neural activation. The evolved robots exhibited various methods of locomotion, including crawling, ratcheting and some forms of pedalism (Figure 2. Selected robots are then replicated into reality: their bodies are first fleshed to accommodate motors and joints, and then copied into material using rapid prototyping technology (Figure 3). Temperature-controlled print head extrudes thermoplastic material layer by layer, so that the arbitrarily evolved morphology emerges pre-assembled as a solid three- dimensional structure without tooling or human intervention. Motors are then snapped in, and the evolved neural network is activated (Figure 4. The robots then perform in reality as they did in simulation.

3.3   Evolving Modularity with Generative Encodings (Tinkerbots)

The main difficulty for the use of evolutionary computation for design is that it is doubtful whether it will reach the high complexities necessary for practical engineering. Since the search space grows exponentially with the size of

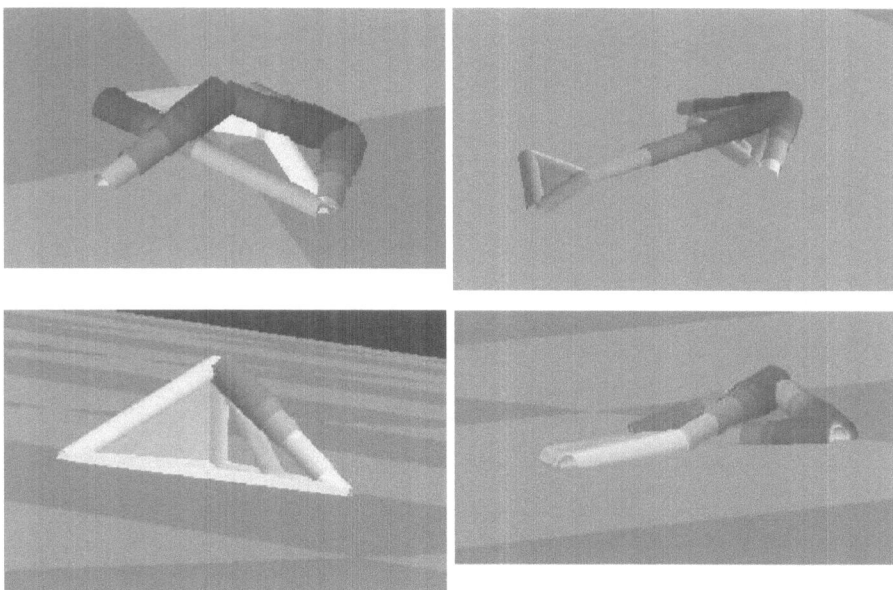

Fig. 2. Several different GOLEM's, evolved in simulation. Brains not shown. See http://demo.cs.brandeis.edu/golem

the problem, search algorithms that use a direct encoding for designs will not scale to large designs. An alternative to a direct encoding is a generative specification, which is a grammatical encoding that specifies how to construct a design, [33,2]. Similar to a computer program, a generative specification can allow the definition of re-usable sub-procedures allowing the design system to scale to more complex designs than can be achieved with a direct encoding. Ideally an automated design system would start with a library of basic parts and would iteratively create new, more complex modules, from ones already in its library. The principle of modularity is well accepted as a general characteristic of design, as it typically promotes decoupling and reduces complexity [36]. In contrast to a design in which every component is unique, a design built with a library of standard modules is more robust and more adaptable [25]. Our system for automated modular design uses Lindenmayer systems (L-systems) as the genotype evolved by the evolutionary algorithm. L-systems are a grammatical rewriting system introduced to model the biological development of multicellular organisms. Rules are applied in parallel to all characters in the string just as cell divisions happen in parallel in multicellular organisms. Complex objects are created by successively replacing parts of a simple object by using the set of rewriting rules. Using this system we have evolved 3D static structures [15], and locomoting mechanisms [14]. Figure 5 shows one of these robots is shown both in simulation and transferred successfully into reality [16].

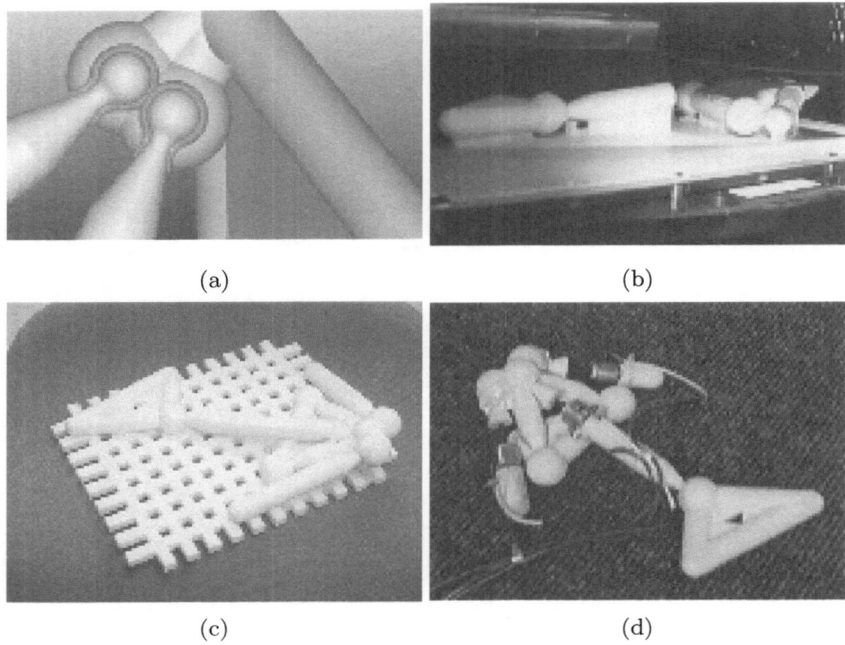

Fig. 3. (a) Fleshed joints, (b) replication progress, (c) pre-assembled robot (d) final robot with assembled motors. See http://demo.cs.brandeis.edu/golem

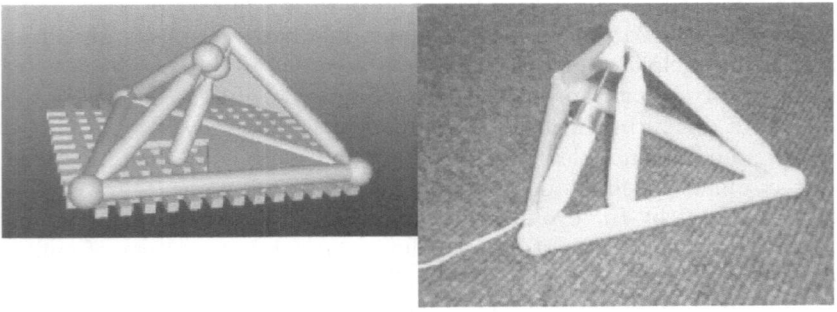

Fig. 4. The geometric representation of a robot, and the actual fabricated machine. Motors are snapped in by hand, and the controller is loaded into a microcontroller chip.

Fig. 5. A virtual "tinkerbot" and the real thing. These robots are evolved with L-systems, and constructed from a fixed collection of parts.

4 Conclusion

Can evolutionary and coevolutionary techniques be applied to real physical systems? In this paper we have presented a selection of our work each of which addresses physical evolutionary substrate in one or more dimensions. We have overviewed research in use of simulations for handling high part-count static structures that are buildable, dynamic electromechanical systems with complex morphology that can be built automatically, and generative encodings as a means for scaling to complex structures. Our long-term vision is that both the morphology and control programs for robots arise directly through morphology and control-software coevolution: starting from primitive controllers attached to primitive bodies the evolutionary system scales to complex, modular creatures by increasing the dictionary of components as stored in the creature encoding. Our current research moves towards the overall goal down multiple interacting paths, where what we learn in one thrust aids the others. We envision the improvement of our hardware-based evolution structures, expanding focus from static buildable structures to buildable robots. We see a path from evolution inside CAD/CAM and buildable simulation, to rapid automatic construction of novel controlled mechanisms, and finally the use of generative encodings to achieve highly complex, modular individuals. We believe such a broad program is the best way to ultimately construct complex autonomous robots whose corporate assemblages consist of simpler, automatically-manufactured parts. Only when robots can economically justify their own existence will they become more ubiquitous outside repetitive industrial use.

Acknowledgements. This research was supported in part by the National Science Foundation (NSF), the office of Naval Research (ONR), and the Defense Advanced Research Projects Agency (DARPA).

References

1. Angeline P. J., Saunders G. M., and Pollack J. B.. An evolutionary algorithm that constructs recurrent networks. *IEEE Transactions on Neural Networks*, 5(1):54–65, 1994.
2. Bentley P. and Kumar S.. Three ways to grow designs: A comparison of embryogenies of an evolutionary design problem. In Banzhaf, Daida, Eiben, Garzon, Honavar, Jakiel, and Smith, editors, *Genetic and Evolutionary Computation Conference*, pages 35–43, 1999.
3. Bentley P. J.. Generic Evolutionary Design of Solid Objects using a Genetic Algorithm. PhD thesis, Division of Computing and Control Systems, School of Engineering, The University of Huddersfield, 1996.
4. Bentley P., editor. *Evolutionary Design by Computers*. Morgan-Kaufmann, San Francisco, 1999.
5. Brooks R.. Intelligence without representation. *Artificial Intelligence*, 47(1–3):139–160, 1991.
6. Cliff D., Harvey I, and Husbands P.. Evolution of visual control systems for robot. In M. Srinivisan and S. Venkatesh, editors, *From Living Eyes to Seeing Machines*. Oxford, 1996.
7. Cliff D., Noble J.. Knowledge-based vision and simple visual machines. Philosophical Transactions of the Royal Society of London: Series B, 352:1165–1175, 1997.
8. Dimos D., Danforth S.C., and Cima M.J.. *Solid freeform and additive fabrication*. In J. A. Floro, editor, 1998.
9. Floreano D. and Mondada F.. Evolution of homing navigation in a real mo-bile robot. *IEEE Transactions on Systems, Man, and Cybernetics*, 1996.
10. Funes P. and Pollack. J.B. Computer evolution of buildable objects. In Phil Husbands and Inman Harvey, editors, *Fourth European Conference on Artificial Life*, pages 358–367, Cambridge, 1997. MIT Press.
11. Funes P. and Pollack. J.B. Evolutionary body building: Adaptive physical designs for robots. *Artificial Life*, 4(4):337–357, 1998.
12. Funes P. and Pollack J. B.. Computer evolution of buildable objects. In Peter Bentley, editor, *Evolutionary Design by Computers*, pages 387 - 403. Morgan-Kaufmann, San Francisco, 1999.
13. Gallagher J. C., Beer R. D., Espenschield K. S., and Quinn R. D.. Application of evolved locomotion controllers to a hexapod robot. *Robotics and Autonomous Systems*, 19:95–103, 1996.
14. Hornby G. S., Lipson H., and Pollack J. B.. Evolution of generative design systems for modular physical robots. In *IEEE International Conference on Robotics and Automation*, 2001.
15. Hornby G.S. and Pollack J. B Body-brain coevolution using l-systems as a generative encoding. In *Genetic and Evolutionary Computation Conference*, 2001.
16. Hornby G.S. and Pollack J. B.. The advantages of generative grammatical encodings for physical design. In *Congress on Evolutionary Computation*, 2001.
17. Husbands P., Germy G., McIlhagga M., and Ives R.. Two applications of genetic algorithms to component design. In T. Fogarty, editor, *Evolutionary Computing*. LNCS 1143, pages 50–61. Springer-Verlag, 1996.
18. Jakobi N.. Evolutionary robotics and the radical envelope of noise hypothesis. *Adaptive Behavior*, 6(1):131–174, 1997.
19. Juillé H. and Pollack J. B.. Dynamics of co-evolutionary learning. In *Proceedings of the Fourth International Conference on Simulation of Adaptive Behavior*, pages 526–534. MIT Press, 1996.

20. Kane C. and Schoenauer M.. Genetic operators for two-dimentional shape optimization. In J.-M. Alliot, E. Lutton, E. Ronald, M. Schoenauer, and D. Snyers, editors, *Artificial Evolution — EA95*. Springer-Verlag, 1995.
21. Kawauchi Y., Inaba M., and Fukuda T.. Genetic evolution and self-organization of cellular robotic system. *JSME Int. J. Series C. (Dynamics, Control, Robotics, Design & Manufacturing)*, 38(3):501–509, 1999.
22. Komosinski M. and Ulatowski. S. Framsticks: Towards a simulation of a nature-like world, creatures and evolution. In Jean-Daniel Nicoud Dario Floreano and Francesco Mondada, editors, *Proceedings of 5th European Conference on Artificial Life (ECAL99)*, volume 1674 of Lecture Notes in Artificial Intelligence, pages 261–265. Springer-Verlag, 1999.
23. Lee W., Hallam J., and Lund H.. A hybrid gp/ga approach for co-evolving controllers and robot bodies to achieve fitness-specified tasks. In *Proceedings of IEEE 3rd International Conference on Evolutionary Computation*, pages 384–389. IEEE Press, 1996.
24. Lipson H and Pollack J.B.. Automatic design and manufacture of robotic lifeforms. *Nature*, 406(6799):974–978, 2000.
25. Lipson, H., Pollack J. B., Suh N. P., 2001, "Promoting Modularity In Evolutionary Design", *Proceedings of DETC'01 2001 ASME Design Engineering Technical Conferences*, September 9–12, 2001, Pittsburgh, Pennsylvania, USA, to appear.
26. Lund H., Hallam J., and Lee W.. Evolving robot morphology. In *Proceedings of IEEE Fourth International Conference on Evolutionary Computation*, pages 197–202. IEEE Press, 1997.
27. Lund H.. Evolving robot control systems. In Alexander, editor, *Proceedings of 1NWGA*. University of Vaasa, 1995.
28. Maja J Mataric and Dave Cliff. Challenges in evolving controllers for physical robots. *Robotics and Autonomous Systems*, 19(1):67–83, 1996.
29. Moravec H.P.. Rise of the robots. *Scientific American*, pages 124–135, December 1999.
30. Morrison K. and Nguyen. T., On-board software for the mars pathfinder micro-rover. In *Proceedings of the Second IAA International Conference on Low-Cost Planetary Missions*. John Hopkins University Applied Physics Laboratory, April 1996.
31. Nilsson N. J.. A mobile automaton: An application of artificial intelligence techniques. In *Proceedings of the International Joint Conference on Artificial Intelligence*, pages 509–520, 1969.
32. Pollack J.B. and Blair A. D.. Coevolution in the successful learning of backgammon strategy. *Machine Learning*, 32:225–240, 1998.
33. Schoenauer M.. Shape representations and evolution schemes. In L. J. Fogel, P. J. Angeline, and T. Back, editors, *Proceedings of the 5th Annual Conference on Evolutionary Programming*. MIT Press, 1996.
34. Sims K. Evolving 3d morphology and behavior by competition. In R. Brooks and P. Maes, editors, *Proceedings 4th Artificial Life Conference*, pages 28–39. MIT Press, 1994.
35. Sincell M.. Physics meets the hideous bog beast. *Science*, 286(5439):398–399, October 1999.
36. Suh N. P.. *The Principles of Design*. Oxford University Press, 1990.

Steps Towards Living Machines

Rodney A. Brooks

MIT Artificial Intelligence Laboratory
545 Technology Square
Cambridge, MA 02139, USA
brooks@ai.mit.edu

Abstract. We still do not understand what it is that makes living matter alive. If we did we could build living machines, but it is clear that we do not have the technology to do that today.

Living machines would be able to self-reproduce, find their own sources of energy, and repair themselves to some degree. They need not necessarily be built from our standard materials, silicon and metal. Living machines will change all of our technologies with equivalent disruption as that introduced by electricity and that by plastics. Living machines will invade the fabric of our everyday lives.

There are three thrusts to trying to build living machines. First is to build robots with partial characteristics of living machines, looking for the key intellectual ideas that make them possible. The second is to use generalized evolutionary systems to investigate possible mechanisms and designs. Generalized evolutionary systems use analogs of physical processes to organize the world for evolving systems, living in that world. The third thrust is to develop a new mathematics of living systems. This new mathematics interacts with the first two thrusts in two ways. It is inspired by the first two thrusts to formalize the notions developed there. Additionally it is used to provide constraints on the design spaces in the first two thrusts, to guide the research work to the appropriate areas.

1   Introduction

We all have an intuitive understanding of when something is alive or not. Of course it is possible to construct boundary cases where it is hard to tell, but most of the time we have very little difficulty in proclaiming that something is living or that it is not living. It is highly unlikely that today we would proclaim any machine as something that is alive.

Despite this we do not have any formal definition of living that captures our folk psychological intuitions. And, we are unable to articulate just what it is about living matter that makes it living, just what it is that differentiates the living from non-living [8].

When we truly understand something we are able to build machines based on the principles we understand. We do not yet know how to build living machines. When we understand the question "What is life?" we will be able to build living

T. Gomi (Ed.): ER 2001, LNCS 2217, pp. 72–93, 2001.

machines — machines that might well be made of inorganic materials, but which will be just as alive as a bacterium, or a dog.

How do we get there from here?

2   Problems

We know that our current robots are not as alive as real living creatures. The six legged robot Genghis [6] can scramble over rough terrain, but it does not have the long term independence that we expect of living systems. The sociable humanoid robot Kismet [5] can engage people in social interactions, eventually people get bored with it and some start to treat it with disdain, as though it is an object, rather than a living creature. The evolution systems for building controllers for robots [28] have not led to arbitrarily competent robots. Something is missing in the robots we build, something that stops us from thinking of them as beyond robot.

Besides robotics, people in ALife try to build systems that exhibit lifelike properties. Ray [30] developed the Tierra system which simulated a simple computer so that he could have complete control over how it worked. Multiple computer programs competed for resources of the processing unit in the simulated computer. Ray placed a single program in a $60,000$ word (his words were only five bits, to approximate the information content in three base pairs of amino acids where there are only twenty proteins coded for plus a little control), memory and let it run. The program tried to copy itself elsewhere in memory and then spawn off a new process to simultaneously run that program. In this way the memory soon filled with simple "creatures", that tried to reproduce by copying themselves.

The code of the computer program was used in two ways. It was interpreted to produce the process of the creature itself, and it was copied to make a child program. The simulated computer, however, was subject to two sources of error. There were "cosmic rays" that would occasionally randomly flip a bit in memory, and there were copying errors, where as a word was being written to memory, a random bit within it might flip. Thus as the memory of the simulated computer filled with copies of the original seed program, mutations appeared. Some programs simply failed to run any more and were removed. Others started to optimize and get slightly smaller as the scheduling policy for multiple programs implicitly had a bias for smaller programs. Before long "parasites", less than half the size of the original seed program evolved. They were not able to copy themselves, but they could trick a larger program into copying them rather than itself. Soon other sorts of creatures joined them in the soup, including hyper-parasites, and social programs which needed each other to reproduce.

When this work was first presented It seemed as though this was the key to building complex life-like systems. Instead of having to be extremely clever, perhaps human engineers could set up a playground for artificial evolution, and interesting complex creatures would evolve. But somehow a glass ceiling was soon hit. Despite many years of further work by Ray and others, and experiments with thousands of computers connected over the Internet, nothing very much more interesting than the results of the first experiments have come along. One hypothesis about this was that the world in which the programs lived was just not complex enough for anything interesting to happen. In terms of information content, the genomes of his programs were four orders of magnitude smaller than the genome of the simplest self sufficient cell. Also the genotype, the coding, and the phenotype, the body, for Ray's creatures were one and the same thing. In all real biology the genotype is a strand of DNA, while the phenotype is the creature that is expressed by that set of genes.

Later, Sims [32] built an evolution system where the genotypes and phenotypes were different. His genotypes were directed graphs with properties that allowed easy specification of symmetry and segmented body parts, such as multisegment limbs. Each element of the phenotype expressed by these genotypes was a rectangloid box that might have sensors in it, actuators to connect it to adjacent rectangloids, and a little neural network, also connected to the neural networks of adjacent body parts. A creature was formed out of many different sized rectangloids, coupled together to form legs, arms, or other body parts. The specified creature was placed in a simulated three dimensional world with full Newtonian physics including gravity, friction, and a viscous fluid filling the volume. By changing the parameters of the fluid it could act like water, air, or a vacuum.

Sims simulated 100 creatures per generation. They were evaluated with a fitness function like how well they swam or crawled, and then the better ones would be allowed to reproduce. As with Ray's system there were various ways in which mutations could happen. Over time the creatures would get better and better at whatever they were being measured for. The first sorts of creatures Sims evolved were those that could swim in water. Over time they would evolve either into long snake like creatures, or into stubbier creatures with flippers.

When these were placed on dry ground they could not locomote very well, but by selecting for creatures that were better at locomoting soon they evolved to be better and better at it. Sims set the creatures a task of trying to grab a green block placed between two of them in a standardized competition format. Many different strategies quickly evolved. Some creatures blindly lashed out and tried to scoop up whatever was in the standard position with a long articulated arm. Others were more defensive, quickly deploying a shield against the opponent, and then grabbing the block at their leisure. Others locomoted towards the block and tried to run off into the distance pushing the block in front of them, scooped in by a couple of stubby arms.

This work re-ignited the dreams inspired by Ray's work that we would be able to mindlessly evolve very intelligent creatures. Again, however, disappointment set in, as there has not been significant improvements on Sims' work in over five years.

Recently Lipson & Pollack [24] implemented a new evolution system with similar capabilities to Sims' program. Besides evaluating their creatures in simulated Newtonian physics they also took the extra step of connecting their creatures directly to rapid prototyping fabrication machines. Their creatures get manufactured physically in plastic, with links and ball joints. A human has to intercede to snap in electric motors in motor holders molded into the plastic. Then the creature is free of cyberspace and is able to locomote in the real world. This new innovation has once again inspired people, but there is still no fundamental new idea which is letting the creatures evolve to do better and better things. The problem is that we do not really know why they do not get better and better, so it is hard to have a good idea to fix the problem.

Both robots and artificial life simulations have come a long way. But they have not taken off by themselves in the ways we have come to expect of biological systems. In Brooks [8], I argued that perhaps something is missing in our understanding of living systems, that perhaps we need to put something more into them so that they will appear and act much more lifelike.

3   What Is Lacking?

We can make a few hypotheses about what is lacking in our robotics and ALife models.

1. We might just be getting a few parameters wrong in all our systems.
2. We might be building all our systems in too simple environments and once we cross a certain complexity threshold everything will work out as we expect.
3. We might simply be lacking enough computer power.
4. We might actually be missing something in our models of biology; there might indeed be some "new stuff" that we need.

The first three of these cases are all somewhat alike, although they refer to distinctly different problems. They are all alike in that if one of them turns out to be true then it means that we do not need anyone to be particularly brilliant to get past our current problems — time and the natural process of science will see us through.

The first case, getting just a few parameters wrong would mean that we have essentially modelled everything correctly, but are just unlucky or ignorant in some minor way. If we would just stumble across the right set of parameters, or perhaps get a little deeper insight into some of the problems so that we could better choose the parameters, then things will start working better. For instance, it could be that our current neural network models will work quantitatively better if we have five layers of artificial neurons, rather than today's standard three layers. Why this should be is not clear, but it is plausible that it might be so. Or it could be that artificial evolution works much better with populations of 100,000 or more, rather than the typical thousand or less. But perhaps these are vain hopes. One would expect that by now someone would have stumbled across a combination of parameters that worked qualitatively better than anything else around.

Now consider the second case. Although there was disappointment in the ultimate fate of the systems evolved by Ray and Sims it was the case that the environments the creatures existed in did not demand much of them. Perhaps they needed more environmental pressure in order to evolve in more interesting ways. Or perhaps we have all the ideas and components that are needed for living breathing robots, but we just have not put them all together all at once. Perhaps we have only operated below some important complexity threshold. While this is an attractive idea, many people have been motivated by the same line of thinking and it has only so far resulted in systems which seem to suffer from "kitchen sink" syndrome. So again, while this is possible I rank it has having low probability of being the only problem.

The third case, that of a lack of computer power, is nothing new. Artificial intelligence and artificial life researchers have perennially complained that they do not have enough computer power for their experiments. However the amount of computer power available has continued to follow Moore's law and doubled every eighteen months or two years since the beginning of each field. So it is getting hard to justify lack of computer power as the thing that is holding up progress. Of course there has been major progress in both fields and many areas of progress have been enabled by the gift of Moore's law. And sometimes we have indeed seen sudden qualitative changes in the way systems seem to function, just because more computer power is available, such as when Deep Blue finally beat Garry Kasparov.

Perhaps our current models of intelligence and life will become intelligent and will come to life if we could only get enough computer power. I am still doubtful that just more computer power, by itself, will be sufficient however.

4   New Stuff

I am putting my intellectual bets on the fourth case, above, that we are still missing something in biology, that there is some "new stuff".

Unlike many who argue that traditional machines are limited, I am betting that the new stuff is something that is already staring us in the nose, and we just have not seen it yet. The essence of the argument that I will make here is that there is something that in some sense is obvious and plain to see, but that we are not seeing it in any of the biological systems that we examine. Since I do not know what it is I can not talk about it directly. Instead, I must resort to a series of analogies.

First consider an analogy from physics, and building physical simulators. Suppose we are trying to model elastic objects falling and colliding. If we did not quite understand physics we might leave out mass as a specifiable attribute of objects. This would be fine for the falling behavior since everything falls in Earth's gravity with an acceleration independent of its mass. So if we implemented that first we might get very encouraged by how well we were doing. But then when we came to implement collisions, no matter how much we tweaked parameters or no matter how hard we computed, the system would just not work correctly. Without mass in there the simulation will never work.

Now I am not suggesting that we need something as revolutionary as mass for our understanding of life. Previous augmentations of existing physical theories have rocked the scientific world. For instance, the discovery of X-rays a century ago which led ultimately to quantum mechanics, or the discovery of the constancy of the speed of light which led to Einstein's theories of relativity both added completely new layers to our understandings of the Universe. We eventually realized that our old understanding of physics was only an approximation of what was really happening in the Universe, useful at the scales at which it had been developed, but dangerously incorrect at other scales.

My version of the "new stuff" is not at all disruptive. It is just some new mathematics, which I have provocatively been referring to as "the juice" for the last few years. But it is not an elixir of life that I hypothesize, and it does not require any new physics to be present in living systems. My hypothesis is that we may simply not be seeing some fundamental mathematical description of what is going on in living systems. Consequently we are leaving out the necessary generative components to produce the processes that give rise to those descriptions as we build our artificial intelligence and artificial life models.

We have seen a number of new mathematical techniques arrive with great fanfare and promise over the last thirty years. They have included *catastrophe theory, chaos theory, dynamical systems, Markov random fields*, and *wavelets* to name a few. Each time researchers noticed ways in which they could be used to describe what was going on inside living systems. There often seemed to be confusion about the use of these mathematical techniques. Having identified these ways of describing what was happening inside biological systems using these techniques researchers often jumped, without comment, to using them as explanative models of how the biological systems were actually operating. They then used the techniques as the foundations of computations that were meant to simulate what was going on inside the biological systems, but without any real evidence that they were good generative models. In any case, none of these ended up providing the breakthrough improvements that the early adopter evangelists had often predicted.

None of these mathematical techniques has the right sort of systems flavor that I am hypothesizing. The closest analogy that I can come up with is that of computation. Now I am not saying that computation is the missing notion, rather that it is an analogy for the sort of notion that I am hypothesizing.

First we can note that computation was not disruptive intellectually, although the consequences of the mathematics that Turing and von Neumann developed did have disruptive technological consequences. A late 19th century mathematician would be able to understand the idea of Turing computability and a von Neumann architecture with a few days instruction. They would then have the fundamentals of modern computation. Nothing would surprise them, or cause them to cry out in intellectual pain as quantum mechanics or relativity would if a physicist from the same era were exposed to them. Computation was a gentle non-disruptive idea, but one which was immensely powerful. I am convinced that there is a similarly, but different, powerful idea that we still need to uncover, and that will give us great explanatory and engineering power for biological systems.

For most of the twentieth century scientists have poked electrodes into living nervous systems and looked for correlations between signals measured and events elsewhere in the creature or its environment. Until the middle of the century this data was compared to notions of control of cybernetics, but in the second half it was compared to computation; how does the living system compute? That has been a driving metaphor in scientific research. I am reminded that early on the nervous system was thought to be a hydrodynamic system, and later a steam engine. When I was a child I had a book which told me that the brain was like a telephone switching network. By the sixties children's books were saying that the brain was like a digital computer, and then it became a massively parallel distributed computer. I have not seen one, but I would not be at all surprised to see a children's book today that said that the brain was like the World Wide Web with all its cross references and correlations. It seems unlikely that we have gotten the metaphor right yet. But we need to figure out the metaphor, and in my view it is likely to be something related to a mathematical formalism of something that we can see all the pieces, but can not yet understand how it fits together.

As another analogy to computation imagine a society that had been isolated for the last hundred years, and that had not invented computers. They did however have electricity and electrical instruments. Now suppose the scientists in this society were given a working computer. Would they be able to reduce it to a theoretical understanding of how it worked — how it kept a data-base, displayed and warped images on the screen or played an audio CD, all with no notion of computation? I suspect that these isolated scientists would have to first invent the notion of computation, perhaps spurred on by correlations of signals that they saw by taking measurements on visible wires, and even inside the microprocessor chip. Once they had the right notion of computation they would be able to make rapid progress in understanding the computer, and ultimately they would be able to build their own, even if they used some different fabrication technology. They would be able to do it because they would understand its principles of operation.

Now we have the analogies for the mathematics that I suspect we are missing. But where might we look for such mathematics? Rosen [31] posits that we may have to generalize our current understanding of physics to understand life, and makes a case for *category theory* being the crucial mathematical tool — I doubt this is correct, but the arguments he makes in the first four chapters of his book are very persuasive that we do need some new mathematical concepts. If not category theory, then what? It is certainly the case that living systems are made up of matter, and that our current computational models of living systems to not adequately capture certain "computational" properties of that matter. Real matter can not be created and destroyed arbitrarily. This is a constraint that is entirely missing from ALife simulations of living systems. Furthermore all matter is doing things all the time that would be incredibly expensive to compute. Molecules are subject to forces from other molecules around them, and physics operates by continually minimizing these forces. That is how the shapes of the membranes of cells come about, how molecules migrate through solution and across barriers, and how large complex molecules fold up on themselves to take the physical shapes that are important for the way they interact with each other as the mechanisms of recognition, binding, or transcription.

But this is just one obvious place to look. The real trick will be to find the non-obvious, for if the juice hypothesis is true, that must be where it is hiding.

It might turn out that for all the different aspects of biology that we model that there is a different juice that we are missing.

For perceptual systems, say, there might be some organizing principle, some mathematical notion that we need in order to understand how animal perceptions systems really work. Once we have discovered this juice we will be able to build computer vision systems that are good at all the things they currently are not good at. These include separating objects from the background, understanding facial expressions, discriminating the living from the non-living and general object recognition. None of our current vision systems can do much at all in any of these areas.

Perhaps other versions of the juice will be necessary to build good explanations of the details of evolution, cognition, consciousness or learning, will be discovered or invented and let those subfields of AI and ALife flower.

Or perhaps, there will be just one mathematical notion, one juice, hat will unify all these fields, revolutionize many aspects of research involving living systems, and enable rapid progress in AI and ALife.

5   Are We Clever Enough?

My version of the "new stuff" argument boils down to wanting us to come up with some clever analysis of living systems, and then to use that analysis in designing better machines. It is possible that there is indeed something beyond such an analysis, that there really is some "new stuff" that relies on different physics than we currently understand. We have seen this happen twice in just over the last one hundred years, first with radiation which ultimately lead to quantum mechanics, and then with relativity. Sometimes there really is new physics, and if we use that term broadly enough, it must surely cover what it is that makes things alive.

The next concern is whether people will ever be clever enough to understand the "new stuff", whether it is my weak version of just being better analysis, or whether it really does turn out to be some fundamentally new physics. What are the limits of human understanding?

Patrick Winston at MIT in his undergraduate artificial intelligence lectures likes to tell a story about a pet raccoon that he had when growing up in Illinois. The raccoon was very dextrous, able to manipulate mechanical fasteners and get to food that was supposed to be inaccessible. But Patrick says that it never occurred to him to wonder whether some day racoonkind might eventually build a robot raccoon, just as capable as themselves. His parable is meant to be a caution to MIT undergraduates that perhaps they are not as smart as they like to think (and they certainly do like to think that way...) and that perhaps we humans have too much hubris in our quest for artificial intelligence.

Could he be right? Perhaps there is some super version of Godel's theorem that says that any life form in the Universe can not be smart enough to understand itself well enough to build a replica of itself through engineering a different technology. It is a bit of a stretch to think about the formal statement of such a theorem, but not difficult to see that in principle it might be the case that such a thing is true.

If such a thing is true it still leaves open the door for someone smarter than us to figure out how we work, and to build a working machine that is just as emotional, and just as clever as us. In what way could someone else be intrinsically smarter than us? This again is a issue of not wanting to be not special.

Let us consider someone who is not quite as smart as all of us, but does not realize it. Woodward *et al* [39] report on a patient, called JT, who suffered a brain hemorrhage. As with many such people JT was left with some mental functions intact, but others impaired. In JT's case he suffered from a *color agnosia*, a disruption of his ability to understand colors. His disruption, however, is most interesting for how mild it is. By comparing JT to normal control subjects Woodward and his coworkers were able to ascertain that JT had normal performance in distinguishing colors from each other. In abstract color gratings he was able to count various bars distinguished only by changes in color and not brightness. So called *color blind* people are very poor at this task. JT was also able to manipulate color words, successfully verbally filling in the color names in phrases like "green with envy". So he is able to see different colors, and use color words. The problem for JT comes when he must associate these two different abilities. When asked to name color patches on a screen JT got only about half of them right, most typically getting similar colors confused — e.g., yellow and orange, or purple and pink. When asked to use color pencils to color in line drawings of familiar object JT was able to get about three quarters of them right. These experiments were carried out, and some were repeated, over a period of years. Thus JT was not improving his abilities, he was permanently damaged.

This seems very strange indeed. Someone who can "see" colors, and can use color words, but can not make the right associations between them. Surely we would all be clever enough to figure things out and simply re-learn the colors? But if that is so, why can JT not? He was a functioning adult with a technical job before his hemorrhage. He was still a functioning adult person afterwards, but had a deficit that the rest of us can see. How many "deficits" do the rest of us have, but that all of us in our recursive ignorance do just not see?

With small wiring changes in their brains, people can have strange deficits in what they can reason and learn about. As the product of evolution it is unlikely that we are completely optimized, especially in cognitive areas. Evolution builds a hodge-podge of capabilities that are adequate for the niche in which a creature survives. It is entirely possible that with a few additional wiring changes in our "normal" brains we would have new found capabilities. These could be new found capabilities that we can not currently reason about, just as with the agnosia patient. They would be capabilities that our own special reasoning, of which we as humans are so proud, are not capable of reasoning about, without the wiring already in place.

As we relate to chimpanzees and macaque monkeys and their intellectual abilities, we can imagine, at least in principle, a race that has evolved that has pretty much all the capabilities of our brains, but with additional wiring in place, and perhaps even with newer modules. Just as some of our modules have capabilities that are not present in chimpanzees, a super race might have modules and capabilities that are not even latently present in us. Rather than being from Earth, perhaps this super race is from another planet, orbiting one of the billions of stars in one of the billions of galaxies that populate the Universe. What will happen this super race looks at us? Will they see a raccoon with dextrous little hands, will they see an impaired agnosia sufferer who simply can not reason about things that are obvious, or will they see a race of individuals capable of building artificial creatures with capabilities similar to their own?

If we are not capable of building instances of ourselves perhaps we can hope to rely on evolutionary techniques. But even there we will still need to understand the "juice" in order to be able to build an open-ended enough system.

6   How to Find the New Stuff

I will use the term *Earthlife* to refer to things that fit under our current understanding of organically-based living entities here on Earth. That, of course, is the only form of life that we currently use the word life to refer to. But any day now we might detect life that is not Earthlife — organic life on other planets or bodies within our solar system, or signs of organic life on other nearby solar systems. Or we might invent new forms of life that are life but are not Earthlife.

I propose a three-pronged attack on the problem of understanding life.

1. Building robots that emulate fundamental aspects of Earthlife.
2. Carrying out large scale computational experiments that explore developmental and evolutionary aspects of Earthlife.
3. Developing mathematical formalisms that are inspired by the first two prongs, and that explain them and predict things about further experiments in those prongs, and predict things about Earthlife.

6.1   Guides

There are a number of points that we should keep in mind.

Exponential growth. Growth that is linear in time is always going to get dominated by growth that is an exponential function of time. Reproduction of living systems generates an exponential process in the absence of resource limitations. The spread of processes within a living system are often exponential. Going after exponential processes, processes that are exponential in their physical growth, can be the key to something dominating its environment. And this applies equally well to molecules dominating the interior of a cell as it does to an organism dominating an ecological niche.

Adaptivity. Evolution has produced systems which are very adaptable at the neural level. Even simple animals like *C. Elegans* with fewer than a thousand cells have hundreds of neurons [38]. Often this is viewed as a sign that the animals are able to adapt to new environments. But perhaps the real strength is that it allows evolution to experiment with details of the animal, and it can adapt to whatever its configuration turns out to be. I.e., it is not adaptability to the environment which may be important, but rather adaptability to mutations. What does this mean for systems without neurons? Perhaps there are other mechanisms in all living systems which provide adaptivity. Neural adaptivity was just a radical form which let evolution really take hold.

Sloppiness. McMullin [27] points out that the cellular automata that we have seen, such as von Neumann's 29 state system, while providing examples of self-reproducing machines, are completely fragile. A single cell with a different state will most likely render the configuration completely impotent. In the case of a Conway game of life configuration it will most likely lead to its complete annihilation. This is in complete contrast to Earthlife. A worm with a few extra cells here or there will still act just as well as a worm without. A cell with 587 copies of a particular protein will probably act pretty indistinguishably from one with 583 copies of a protein. There is a sloppiness in biological systems that gives room for variation without loss of function or robustness. And that sloppiness is naturally accommodated and utilized.

Phase changes. Langton [23] demonstrates that a sudden qualitative change can happen in the behavior of cellular automata as, by at least one measure, the so called λ parameter, as a small quantitative change is made in the behavior of the transition rule. This is like a phase change in a physical system, where at a critical temperature a very small change will convert a liquid to a solid, for example. There may well be many, many, instances of this sort of phenomenon all across biological and complex computational systems. One might expect it in metabolic systems, and in evolutionary systems for example.

Life effects its environment. A key aspect of living systems is that they are situated in the world in a very dynamic way. In the past *situated* has often been used to refer only to the computational implications [7]. However biological creatures, and their symbiotic cohorts, used their situatedness in a deeper way. They individually, and together, perhaps spread out over both time and space, modify their environment in a way which supports them or their distant progeny. Life reshapes the non-living environment (and indeed this will be the only way in which we will be able to detect life on extra-solar planets for the foreseeable future). The reshapen environment supports life, whether it be the life that reshaped it, or morphogenically different life at some future time. Life does not exist just in its *ambiance* [31]. Life and its ambiance are coupled together in deep and historical ways.

Darwinian evolution. Darwinian evolution is both a self fulfilling prophecy and an emergent property of certain systems (just what systems is a difficult question

I think). It is not imposed from above. Not all artificial evolution systems are Darwinian. Ray [30], for instance, defines a reaper that operates from above, outside of the dynamics of the self-producing programs that he has. This puts a severe limitation on what can evolve. Better forms of evolution cannot evolve because the system is not open ended. Earthlife systems on the other hand operate in an open ended environment. Things survive because they do. Things do not survive because they do not do as well as the ones that do. If something figures out a new trick, its *exponential* power will soon let it dominate some ecological niche.

6.2   Directions

Some directions that might yield an understanding of life, and therefore the "new stuff" or "juice" are as follows:

Robots. Getting away from silicon and steel to new materials, self repairing robots, energy self sufficient robots, self reproducing robots.

Computational experiments. Pre-biotic chemical self-organization, evolution of DNA and RNA based coding, evolution of the cell, adaptivity of neural systems during development, evolution of higher level capabilities.

Mathematics. A way of thinking, formalisms, and theorems. These would be inspired by the building of robots and of computational experiments.

7   Example Research Projects

In the following section we give two example experiments, one with robots and one with purely computation which may lead to an understanding of life and therefore allow us to ultimately build better robots.

7.1   Neural Control

[I am indebted to Tony Prescott for introducing me to some of the literature cited in this section.]

The control systems we build for our robots are not as robust as those of real animals. Often people use neural networks to control robots, but these networks are abstractions of real networks. These abstractions were primarily developed to model classification and pattern recognition. If we look at simple animals we find that their neural networks are not doing such classifications, nor are they in any sense information processing systems. Neural networks probably evolved for motor control originally, and by returning to those roots, perhaps we will find better ways to control our robots. This section talks about some of the inspirational issues found by looking at very simple animals. We should attempt to build robotic models of similarly primitive animals in order to really understand how neural systems work.

The earliest models of neurons, those of McCulloch & Pitts [26] made an abstraction of them as on-off devices. Signals between neurons are on-off signals, and the whole system is synchronously clocked into uniform time segments. In the immediately following paper in the same issue of the journal they along with another author, Landahl *et al* [22] make an argument about the equivalence between their on-off neurons and those that communicate by spike trains using a probabilistic argument. Later von Neumann [36] tackled the same problem with more generality and with his typical tenacity. Both sets of results rely on treating just the instantaneous spike train rate as being all that is important, and both model neurons as having uniform dendritic trees and neural tissue synchronous networks where all connections have identical propagation times.

Interestingly, at other times, von Neumann [35] did point out the differences between the "formal neural networks" of McCulloch and Pitts, and the real neural networks of biological systems. In particular he was frustrated by the tolerance to local errors of the latter, and fragility of the former to such local errors (he estimated that in a digital computer a single bit error in about 30% of all operations would lead to total failure of the program).

These early formalisms in slightly modified forms have remained with the field of computational neural systems for almost sixty year. The signals on the connections between neurons are now often real valued, but the implicit synchronous timing of the network is still there. That implicit timing makes it hard to deal with recurrent networks as the dynamics of loops are very much undercut by the consistent clocking of the whole system.

The innovations in formal neural networks have been to put learning in weights on these links, weights that are multiplied by the incoming signal strength at the time of synchronized firing. There is very strong biological evidence for this being a somewhat realistic model of what happens at the synapses in real nervous systems. But these models hardly begin to capture the richness of real neurons. Some work has been done to rectify this, but mostly only by people working on isolated neurons, people who are much more interested in it for science's sake of understanding neurons, rather than building computational systems modeled on neurons. In general computationalists have not gone after realistic models, although there have been some attempts. The leaky integrator approach to modeling spike trains assumes that the real number on the connection is the instantaneous rate of spiking. But again recurrent networks and spike dynamics are not attended to. One can see Arbib [3] for details of many different models. Quartz & Sejnowski [29] have made similar complaints to ours and suggested other ways to tackle these problems.

Recently formal neural networks have been most strongly associated with learning, and in particular very formal models of learning from the machine learning community. This puts a neural network into the framework of being a function. This framework outlaws the rich dynamics that are seen in real neural networks as they control a creature. There is much richness to be mined by considering a bigger question of how real neural networks are part of a complex dynamics of behavior control. There still might be much to learn from real neural networks.

In the following sections we examine three very primitive organisms to get inspiration on what neural networks might really be doing.

Hydroid polyps. One of the most primitive forms of animals is the hydroid polyp that swims by moving its tentacles. These seem to be coordinated electrically without neurons and axons, but via electrical conduction between more conventional cells [15].

From this, and other evidence from siphonophores and even Drosophila (which have non-neural electrical connections between cells in their salivary glands) it appears that intercellular electrical conduction between cells was present before specialized neurons arose. The functional effect of these early forms of conduction often was to synchronize contractions over remote areas of an organism.

In addition to non-neuronal transmission for locomotion, Hydra also have a very primitive neural system for feeding behaviors [25]. It is interesting because of its primitiveness in two ways. First, all its neurons have sensory inputs, but can have outputs both to other neurons and to effectors. This is very reminiscent of details of the robot Herbert built by Connell [10] — it had even more restrictive connectivity constraints resulting in a very primitive behavior controller, but was nevertheless able to produce quite sophisticated behavior. Second, the hydra neural system is always in a state of flux. Neurons are lost at the extremities and are replaced further upstream by morphological transformations of other forms of cells. The nervous system is thus somewhat like a glacier — replaced at its core as it flakes off at its extremities. Additionally it seems that the neuronal cells play additional roles in the regulation of growth and production of chemical gradients. This should not be surprising if the neurons have only just been co-opted into their now traditional role — vestiges of their former purpose are likely to abound.

Modelling a hydroid polyp will confront a number of interesting aspects of neural system controlling robots. First the development of the neural system can not be ignored as the neural systems of hydra are always in development. Second, since it is in a state of flux, the network cannot become overly complex, yet it must support the rich behavior that the polyp exhibits — how does it develop, and what reinforcement signals (if that is the right way of thinking about it) direct the development of the neural system?

Jellyfish. Horridge [14] describes the neural system of an immature jellyfish Aurellia. The creature is radially symmetric with a number of arms and a mouth in the middle. When swimming all the arms move in synchrony in a pulsating manner, expelling water from the volume between them. When feeding, an arm that touches food sweeps towards the mouth and the mouth turns towards it. Multiple arms can fire in rapid succession so that it is simultaneously feeding on many food particles. Horridge demonstrates through a series of nerve fiber incisions that there are two distinct nerve nets with very different arrangements. One net made up of giant fibers is concerned with a radial muscle that makes the arms beat together. Another net, called the diffuse nerve-net, is much more extensive over the body of the jellyfish. Anatomically this net resembles that of the more primitive hydroid polyps. It is responsible for the feeding behaviors of individual arms. All its neurons have very tight connections with sensory cells.

The two networks of this jellyfish are thus correlated with its two main behaviors — swimming and feeding. It is worth noting that the neural networks are nevertheless quite sophisticated. In particular there is the problem of how to get all of the circular muscle to contract simultaneously. In bilaterally symmetric animals like squid with a central brain, faster nerves are used to excite more distant parts of the body so that a synchronized full body response can occur. In a radially symmetric jellyfish, however, the excitation for a swimming pulsed water jet can occur anywhere in a 360 degree range. Having connections of appropriate lengths from each point to everywhere else would lead to a combinatorial explosion of nerve cells. The jellyfish must find another way to communicate quickly with the distal parts of the muscle.

Mackie [25] reports that the jellyfish Aglantha solves the problem in a brute force manner by having a single very fast giant axon that spreads the excitation all the way around the ring in 2 to 4 milliseconds. The bell contraction is thus effectively simultaneous.

Spencer *et al* [33] deduced that the jellyfish Polyorchis uses a very different approach. There is a ring of motor neurons around the muscle. A spike that initiates at one point and propagates around the neural ring changes shape as it travels, getting less intense and somewhat narrower. The muscles respond to spikes with a latency that is negatively correlated with the intensity of the spike. Thus a part of the muscle near the stimulation gets the spike before a more distant part of the muscle, but it waits to respond to it longer, giving the distant part of the muscle time to get stimulated. In this case the information content of the spike is very much a function of how far it has traveled. The metaphor of a sender broadcasting a message that is received at various receivers completely breaks down; the message itself changes as it is transmitted, and the changes result in different behavior at each receiver.

The swimming behavior of jellyfish is controlled essentially by an oscillator. All the interesting parts of neural networks, traditionally ignored by computationalists, are in service of making this oscillator work — not in service of information processing. In particular any robotic model of a jellyfish that tries to have realistic neural networks will need need to model spike dynamics, and deal with time dependent computations done in dendritic trees.

Polyclad flatworms. Polyclad flatworms provide a level up from jellyfish where we might hope to uncover some of the fundamental ways in which brains work. They do have a brain, ranging in size from a few hundred neurons to less than two thousand, where all the neural systems are integrated, and it is perhaps the simplest brain that can be studied. Furthermore the animals do not die if their brain is removed and so its role in their behavior can easily be studied — e.g., Koopowitz [18] first used this technique to show the role of the brain in the complex feeding behavior of Planocera gilchristi.

Koopowitz & Keenan [21] later gave an overview of another flatworm with similar feeding behaviors, Notoplana acticola. They argue that the role "of the brain is to regulate the operation of peripherally positioned local reflexes". They base this on comparison of the behaviors of animals with intact brains and those with their brains removed.

The brain of notaplana has twelve large nerve connections on its dorsal (lower) layer. Eight of these connections, splice the brain into four longitudinal nerve cords along the body of the animal. The brain is about 0.6mm in diameter, and the full creature is about 25mm long.

A normal animal responds to a dead brine shrimp touching it on the side of the animal at the rear by first grasping it with its body. It then turns the front of its body to that side, regrasps the shrimp and transfers it to its mouth roughly in the middle of the body. After eating two or three shrimp the flatworm is usually satiated and accepts no more food.

When the brain is removed the front of the body no longer turns towards the shrimp. Instead the rear side of the body that does the initial grasp transfers the food directly to the mouth. Furthermore the flatworm does not get satiated but continues to eat. If the connections from the brain to one side of the body, the right side, say, are removed then the left side will still show satiation and normal feeding, but the right side will not show either. After a few days however the right side will start to show the normal behavior again, even if the connections on the right side are not allowed to heel. New connections via the left side are involved in this reappearance, because now if just the left side connections are severed both sides of the animal show the decerebrate behavior. The same thing happens if the whole right half of the brain is removed. There is a recovery of function, although no regrowth of the brain is observed.

Because of this experiment, Koopowitz and Keenan see the brain as an inhibitory device. Rather than arguing that the brain evolved as a control mechanism situated close to the animals' sensors in the head, they propose that the flatworm needed to evolve its brain in order to inhibit behaviors. Because of bilateral symmetry, the behaviors on one side needed to be inhibited to avoid conflict with an ongoing behavior on the other.

Davies et al [11] summarize a large number of experiments that remove the brain of Notoplana acticola, transplant it to another creature and reseat it in new geometric configurations.

After brain transplants the creatures were tested for four behaviors over a period of days:

1. The ability of the animal to right itself after being flipped over.
2. The ability to locomote.
3. The ability to withdraw and locomote away from a noxious stimulus.
4. The normalcy of feeding behaviors.

The brains where cut from creatures all the way through their body. They were then placed in the body of recipient animals (themselves the brain donor for their brain donors) in one of four orientations. The brains were flipped or not in the front/back direction (or anterior/posterior direction, A/P), or in the up/down direction (or dorsal/ventral direction, D/V).

The animals with brains inserted in the normal orientation or in the A/P inverted orientation all did very well on all four tests after 20 days. It is interesting to note that the A/P inverted individuals did walk backwards more often than non A/P inverted animals during their recovery time.

The righting behavior was recovered rapidly by all animals except those with just D/V inversion — they took longer. The ones with D/V and A/P inversion righted themselves much sooner. Likewise the D/V only inverted animals did not recover avoidance behaviors at all well, although D/V and A/P inverted animals did.

Lastly all animals recovered feeding behaviors to some extent. The normal orientation animals almost all recovered completely. The D/V and A/P inverted animals did rather poorly, and the either D/V or A/P animals recovered moderately well.

Koopowitz [19] reports on further experiments where the brains were rotated 90 degrees. In that case brain dominated behavior does not reappear. When the brains were inserted down the center line of a recipient animal, but further to the rear they were also able to regain brain dominated behavior. He also describes the neuroanatomy, neurophysiology, and neurochemistry of the polyclad flatworms. He concludes that they include most of the neural aspects of higher creatures, although of course the latter have probably evolved more complex circuits using the same basic components. There are more details of the complexity of Notoplana neurons to be found in Keenan *et al* [17] and Koopowitz *et al* [20].

The remarkable plasticity of flatworms following brain transplantation opens up a number of interesting issues. Clearly the symmetry of the brain is importantly preserved in some way in the successful transplants. And clearly the longitudinal nerves in the body of the flatworm are important for tolerance to brain misplacement and to one-sided brain removal. But how do the twelve main nerves from the brain get the right signals to connect to the appropriate peripheral nerves? And how do the contra-lateral peripheral nerves get the right signals (if indeed this is an appropriate question) to reconnect when their half of the brain has been removed? And how does a half brain learn how to operate the other side of the body, or how to reverse its internal walking sequencing? The underlying question that all of these questions skirt around is what is the nature of what is going on within the transplanted brains? The natural way of phrasing this question is what are the computations going on in the transplanted brains? But perhaps that is prejudicing the question. Perhaps there are better metaphors to use.

Building a robot with the plasticity of Notoplana will be very challenging. Imagine being able to unplug its main processor chip and rotate in the socket and have the robot adapt and continue to function.

7.2   Pre-biotic Life

How did life begin? Let us assume it began on Earth rather than being seeded by a bacteria carrying meteorite. Even if that was the case then ultimately life did first begin somewhere else in the Universe, and we may as well assume it was on a planet, since we know nothing to the contrary, and we may as well assume that it did occur on Earth, since at this writing we know nothing to the contrary.

So the problem is how non-living matter on Earth organized itself to produce living matter. There was no external guiding hand. It had to come from the physics, and then chemistry, of the pre-biotic soup of Earth, along with the physical properties of the Earth at the time, or over the relevant time period.

Wills & Bada [37] give an outline of how this may have happened and their chapters five through seven are of particular interest, as they lead through the steps that could get molecules organized to a point just a little pre-life.

They first hypothesize a world where small organic molecules, including amino acids, are produced randomly by purely physical processes. There is ample evidence that this could have happened in the early history of our Earth. Chapter five outlines the role of physical processes such as day-night cycle, tides, and storms, and the effect of rocks as anchor points, in providing a sorting and non-uniform distribution of these molecules in tide pools and other places. This is the beginning of self organization.

Chapter six analyzes different sorts of chain molecules, the well known DNA and RNA, along with other slightly simpler molecules, such as PNA (P for Peptide in contrast to R for ribose) that support the same set of bases G, A, C, and U, with which we are all familiar. They show how fragments of such molecules can get spontaneously copied in the right non-living medium, and how some RNA chains act as catalysts for such copying. The key suggestion is that Darwinian selection could act in this world even though it is not living — once a self replicating molecule arises it will quickly proliferate. This is an example of the power of exponential processes.

Chapter seven then speculates how a genetic code could have arisen on top of these chains. First with very simple coding for hydrophobic versus water loving amino acids, producing different proteins on each transcription, but with grossly similar chemical properties. Within that world then the basic molecular structures energy utilization may have come into play — again spurred on by Darwinian effects even in this pre-cellular pre-life world. These physically isolated groups of molecules, clinging to rocks, would then ultimately have evolved into archaea and bacteria.

Can we build a simulation where the same sort of thing happens? Where non-living physics gives rise to computationally living entities within a large scale physical simulation. If we could do this, and have life arise spontaneously we would have an experimental system where we could really get at the issue of living versus non-living. It would perhaps gives us some insights into appropriate formalisms, and whether there is some sort of as yet unknown organizational principles in living systems.

I have clearly hedged here on the definition of *living*, trying to have my cake (or not) and eat it too (or not). I am saying that if we build a simulation system where life can arise spontaneously then we will have a place to study the difference between living and non-living. I think I mean that if we can build a computational simulation system where things that are *unmistakenly* living to a reasonable observer, then we will have two stakes in the ground, at non-living and at an instance of living, where we will be able to look at all the gradations along the way, and hopefully we will then be able to attain enlightenment.

The problem then is how to build a holodeck for molecules, chains, proteins, lipids, and bacteria or archaea.

The main question is what scale of abstraction should our computational experiment have?

There are at least three orthogonal parts to the answer to this question.

The first part of the answer to this question is whether the experiment should be physics-based or bulk property based. In the first case we would have a simulation of individual elementary objects, moving around in some n-dimensional space obeying some laws of physics and chemistry. Everything would emerge from that. In the second case we would have statistical approximations, e.g., solutions with certain concentrations, rather than simulated individual objects floating about in simulated water. The first case has not really been done very much, and the best analogy is the work with cellular automata. There is a real problem in cellular automata about how to capture the permanence of matter, although they have been early attempts [2]. The second case has been used often, for instance by Takagi *et al* [34].

The second part of the answer to this question is what level of individual elementary objects should be used. Some options include subatomic particles, atomic particles, simple molecules (amino acids, riboses, etc.), complex proteins, or some sort of computational element — e.g., lambda expressions.

The third part of the answer to this question is whether the real chemistry, like that of carbon and peptides, should be simulated, or whether some artificial chemistry which has interesting properties should be chosen/tweaked/tuned.

My current guess is that the most promising/interesting approach is to do a physics based system, with artificial chemistry, and either atomic or small molecule elements.

The chemistry needs to be rich enough to support many forms of interaction, and a massive combinatorics so that interesting behaviors can arise without too many elements coming together.

It needs to be able to give rise to chains, base pairs, codons, transcriptase, proteins, energy pumping systems, etc.

Then there are some independent questions, whether or not the above option is chosen.

What should the dimensionality n be? Clearly $n = 1$ is completely uninteresting. Most of the work in cellular automata has been with $n = 2$, although there has been some work for $n > 2$. In Abbott's *Flatland* [1] one sees the inherent difficulties in working in two dimensions only. In fact in cellular automata with complex constructions one sees that most of the area of the construction goes towards crossing signal paths in ways that they do not interfere [9]. Since Earthlife appears to be $n = 3$, it seems reasonable to try for that. Of course it might turn out that everything is much easier in $n > 3$. Perhaps the Universe is teaming with $n > 3$ life and we don't notice it because we and Earthlife are an $n = 3$ anomaly...

Earthlife works because of water and solubility. Should a simulation even at an atomic or small molecule level handle solubility through simulation of individual solvent elements. That seems unlikely. It would be just too computationally expensive. We probably need to have a special "ether" which plays the role of water, but uses a very different basis for its simulation. Should it have bulk properties? Should it be subject to gravitation? How do we handle temperature?

All these are difficult questions and the temptation is to simplify away the "water". But then we may lose opportunities to exploit micro-properties that we may not currently understand well — like Tom Ray lost a lot of power of Darwinian evolution by having the reaper in Tierra.

Earthlife also made use of other non-organic parts of its early environment, such as rocks as anchoring points. These two could be simulated atomically, but most likely they should be handled in some meta-level way as with the "water".

Once a method of simulation has been determined there are many rich sources that can guide how the systems will be put together. Gesteland *et al* [13] is a very recent review of the evidence for, and arguments for, the existence of a form of life that came before our current DNA based life. The steps towards that RNA world are discussed by Joyce & Orgel [16]. Once that is in place there are many ideas about the evolution of genetic codes and the interplay with the structure of amino acids and proteins [12], [4], [34].

8   Conclusion

In this paper I have argued that there are still some fundamental questions about the nature of living systems that are holding up the full potential of evolutionary robotics. I have made some suggestions about how we might tackle these problems through building robots and carrying out computational experiments.

References

1. Edwin A. Abbott. *Flatland: A Romance of Many Dimensions*. Dover Publications, Inc., New York, NY, 1884. 6th Edition, Revised with Introduction by Banesh Hoffmann, 1952.
2. Michael A. Arbib. Simple self-reproducing universal automata. *Information and Control*, 9:177–189, 1966.
3. Michael A. Arbib, editor. *The Handbook of Brain Theory and Neural Networks*. The MIT Press, Cambridge, Massachusetts, 1995.
4. Steven A. Benner, Petra Burgstaller, Thomas R. Battersby, and Simona Jurczyk. Did the RNA world exploit an expanded genetic alphabet? In Raymond F. Gesteland, Thomas R. Cech, and John F. Atkins, editors, *The RNA World*, pages 163–181. Cold Spring Harbor Laboratory Press, Cold Spring Harbor, New York, second edition, 1999.

5. Cynthia L. Breazeal. *Building Sociable Robots*. MIT Press, Cambridge, Massachusetts, 2001.
6. Rodney A. Brooks. A robot that walks: Emergent behavior from a carefully evolved network. *Neural Computation*, 1(2):253–262, 1989.
7. Rodney A. Brooks. New approaches to robotics. *Science*, 253:1227–1232, 1991.
8. Rodney A. Brooks. The relationship between matter and life. *Nature*, 409:409–411, 18 January 2001.
9. E. F. Codd. *Cellular automata*. ACM Monograph Series. Academic Press, New York, 1968.
10. Jonathan H. Connell. *Minimalist Mobil Robotics: A Colony-style Architecture for an Artificial Creature*, volume 5 of *Perspectives in Artificial Intelligence*. Academic Press, Inc., Boston, Massachusetts, 1990.
11. Lynnae Davies, Larry Keenan, and Harold Koopowitz. Nerve repair and behavioral recovery following brain transplantation in notoplana acti cola, a polyclad flatworm. *The Journal of Experimental Zoology*, 235:157–173, 1985.
12. Stephen J. Freeland and Laurence D. Hurst. The genetic code is one in a million. *Journal of Molecular Evolution*, 47:238–248, 1998.
13. Raymond F. Gesteland, Thomas R. Cech, and John F. Atkins, editors. *The RNA World: The Nature of Modern RNA Suggests a Prebiotic RNA*. Cold Spring Harbor Laboratory Press, Cold Spring Harbor, NY, second edition edition, 1999.
14. Adrian Horridge. The nervous system of the ephyra larva of aurellia aurita. *Quarterly Journal of Microscopical Science*, 97:59–74, March 1956.
15. G. A. Horridge. The origins of the nervous system. In Geoffrey H. Bourne, editor, *The Structure and Function of Nervous Tissue*, volume 1, pages 1–31. Academic Press, New York, 1968.
16. Gerald F. Joyce and Leslie E. Orgel. Prospects for understanding the origin of the RNA world. In Raymond F. Gesteland, Thomas R. Cech, and John F. Atkins, editors, *The RNA World*, pages 49–77. Cold Spring Harbor Laboratory Press, Cold Spring Harbor, New York, second edition, 1999.
17. C. Larry Keenan, Richard Coss, and Harold Koopowitz. Cytoarchitecture of primitive brains: Golgi studies in flatworms. *Journal of Comparative Neurology*, 195:697–716, 1981.
18. Harold Koopowitz. Feeding behaviour and the role of the brain in the polyclad flatworm, planocera gilchrist. *Animal Behaviour*, 18:31–35, 1970.
19. Harold Koopowitz. Polyclad neurobiology and the evolution of central nervous systems. In Peter A.V. Anderson, editor, *Evolution of the First Nervous systems*, pages 315–327. Plenum Press, New York, 1989.
20. Harold Koopowitz, Mark Elvin, and Larry Keenan *In vivo* visualization of living flatworm neurons using lucifer yellow intracellular injections. *Journal of Neuroscience Methods*, 69:83–89, October 1996.
21. Harold Koopowitz and Larry Keenan. The primitive brains of platyhelminthes. *Trends in Neurosciences*, 5(3):77–79, 1982.
22. H. D. Landahl, Warren S. McCulloch, and Walter Pitts. A statistical consequence of the logical calculus of nervous nets. *Bulletin of Mathematical Biophysics*, 5:135–137, 1943.
23. Chris G. Langton. Emergent computation. In Stephanie Forrest, editor, *Computation at the Edge of Chaos*, pages 12–37. MIT Press, Cambridge, Massachusetts, 1991.
24. Hod Lipson and Jordan B. Pollack. Automatic design and manufacture of robotics lifeforms. *Nature*, 406:974–978, 2000.

25. G. O. Mackie. The elementary nervous system revisited. *American Zoologist*, 30:907–920, 1990.

26. Warren S. McCulloch and Walter Pitts. A logical calculus of the ideas immanent in nervous activity. *Bulletin of Mathematical Biophysics*, 5:115–133, 1943.

27. Barry McMullin. John von Neumann and the evolutionary growth of complexity: Looking backward, looking forward... *Artificial Life*, 6:347–361, 2000.

28. Stefano Nolfi and Dario Floreano. Learning and evolution. *Autonomous Robots*, 7:89–113, 1999.

29. Steven R. Quartz and Terrence J. Sejnowski. The neural basis of cognitive development: A constructivist manifesto. *The Behavioral and Brain Sciences*, 20:537–596, 1997.

30. Thomas S. Ray. An approach to the synthesis of life. In J. Doyne Farmer Christopher G. Langton, Charles Taylor and Steen Rasmussen, editors, *Proceedings of Artificial Life, II*, pages 371–408. Addison-Wesley, 1990. Appeared 1991.

31. Robert Rosen. *Life Itself: A Comprehensive Inquiry Into the Nature, Origin, and Fabrication of Life*. Columbia University Press, New York, 1991.

32. Karl Sims. Evolving 3d morphology and behavior by competition. In Rodney A. Brooks and Pattie Maes, editors, *Artificial Life IV: Proceedings of the Fourth International Workshop on the Synthesis and Simulation of Living Systems*, pages 28–39. MIT Press, Cambridge, Massachusetts, 1994.

33. Andrew N. Spencer, Jan Pryzsiezniak, Juan Acosta-Urquidi, and Trent A. Basarsky. Presynaptic spike broadening reduces junctional potential amplitude. *Nature*, 340:636–638, 24 August 1989.

34. Hiroaki Takagi, Kunihiko Kaneko, and Tetsuya Yomo. Evolution of genetic codes through isologous diversification of cellular states. *Artificial Life*, 6:283–305, 2000.

35. John von Neumann. The general and logical theory of automata. In L.A. Jeffress, editor, *Cerebral Mechanisms in Behavior-The Hixon Symposium*, pages 1–31. John Wiley & Sons, New York, NY, 1951.

36. John von Neumann. Probabilistic logics and the synthesis of reliable organisms from unreliable component s. In Claude Elwood Shannon and John McCarthy, editors, *Automata Studies*, pages 43–98. Princeton University Press, Princeton, NJ, 1956.

37. Christopher Wills and Jeffrey Bada. *The Spark Of Life*. Perseus Publishing, Cambridge, Massachusetts, 2000.

38. William B. Wood and the Community of C. elegans Researchers, editors. *The Nematode Caenorhabditis Elegans*. Cold Spring Harbor Laboratory Press, Cold Spring Harbor, NY, 1988.

39. Todd S. Woodward, Mike J. Dixon, Kathy T. Mullen, Karin M. Christensen, and Daniel N. Bub. Analysis of errors in color agnosia: A single-case study. *Neurocase*, 5:95–108, 1999.

Artificial Evolution: A Continuing SAGA

Inman Harvey

Centre for the Study of Evolution
Centre for Computational Neuroscience and Robotics
School of Cognitive and Computing Sciences
University of Sussex
Brighton BN1 9QH, UK
inmanh@cogs.susx.ac.uk

Abstract. I start with a basic tutorial on Artificial Evolution, and then show the simplest possible way of implementing this with the Microbial Genetic Algorithm. I then discuss some shortcomings in many of the basic assumptions of the orthodox Genetic Algorithm (GA) community, and give a rather different perspective. The basic principles of SAGA (Species Adaptation GAs) will be outlined, and the concept of Neutral Networks, pathways of level fitness through a fitness landscape will be introduced. A practical example will demonstrate the relevance of this.

1   Artificial Evolution

Every day we come across sophisticated, highly-tuned machinery that is far far more complex than human designers could begin to imagine designing by standard techniques. I am referring to the animals (including other humans), plants and other organisms that we live amongst. These are self-regulating, mostly self-repairing, self-sustaining machines that can cope with changing situations in an incredibly flexible and adaptive fashion. Their designs are the product of billions of years of natural Darwinian evolution.

Proponents of Artificial Evolution aim to capture and exploit the core parts of this natural design methodology, and use it to design artificial complex systems to have comparable properties: adaptive and robust robot control systems, self-repairing electronic circuits, telecommunications networks that grow and rearrange themselves around disruptions, pharmaceutical drug molecules that match up with a range of targets. We do not have the resources of billions of years of experimentation on one or more planets that Natural Evolution has had, so we must be as efficient as possible, and learn the crucial tricks that Nature can show us.

The context of evolution is a population (of organisms, objects, agents ...) that survive for a limited time (usually) and then die. Some produce offspring for succeeding generations, the 'fitter' ones tend to produce more than the less fit. Over many generations, the make-up of the population changes. Without the need for any individual to change, the 'species' changes, in some sense adapts to the prevailing conditions. There are three basic requirements for Darwinian evolution by Natural Selection:

T. Gomi (Ed.): ER 2001, LNCS 2217, pp. 94–109, 2001.

1. **Heredity**: Offspring are (roughly) identical to their parents ...
2. **Variation**: ... except that they are not exactly the same
3. **Selection**: The 'fitter' ones are more likely to have more offspring than the 'unfit' ones

Variation is usually random and undirected, whereas Selection is usually non-random and in some sense directed. In the natural world, direction does not imply a conscious director. Rather, it reflects the fact that those organisms that are not as well designed for their particular ecological niche as their conspecifics will be less likely to survive and have offspring; the others thereby automatically qualify as 'fitter' for that particular niche — whatever that niche might be. If antelopes are often chased by lions, then it is reasonable to talk of Selection providing a selective pressure for a population of antelope to increase their speed over successive generations, other things being equal.

In Artificial Evolution, unlike Natural Evolution, the human experimenter decides what is going to count as 'fit', in what direction Selection should alter the population over generations. In this sense it resembles agricultural practice, where for thousands of years farmers have been selecting the cows that produce more milk, the crops that produce more grain, and breeding from them for future generations. Even without necessarily understanding the genetic details, the manipulation of DNA underlying the process, farmers have long implicitly understood the basic principles of Heredity, Variation and Selection sufficiently well to improve their crops over the centuries.

2   DNA

As we now know (but Darwin did not), a core mechanism underlying Heredity and Variation is the DNA that we (and other organisms) inherit from our parents and pass on to our offspring. DNA is often treated as though it is a 'blueprint', or a set of instructions setting out how an organism will develop from an initial single cell. Many biologists would say that this view of DNA is in important respects misleading; however, in Artificial Evolution, where we can pick and choose those biological ideas that suit us regardless of whether they give the whole biological picture, we typically do indeed take this simplistic view of Artificial DNA as a blueprint. The crucial aspects of DNA that we borrow for our own purposes are:

1. DNA can be treated as a meaningless string of symbols — Cs Gs As and Ts in humans, perhaps 0s and 1s in a Genetic Algorithm (GA) — that are just mindlessly copied to provide Heredity; perhaps with occasional copying errors to provide Variation.
2. The genotype, the full sum of DNA that an organism inherits, has a crucial role in determining the phenotype, the form and the physical and behavioural traits of an organism.

So to give a simple illustration of Artificial Evolution applied to finding a good design for a paper glider, one could invent a set of symbols that specified how a piece of paper, initially square, is folded. For example, A could mean 'fold the paper towards you about a vertical line through the middle'; B could mean 'fold the paper away from you about a diagonal line from NE to SW'. An appropriate set of such symbols could cover all the possible standard folding moves, and any particular list of such symbols, e.g. GABKJNPD, can be used in each of the two ways listed above: firstly, as a string of symbols that can be mindlessly copied and passed on, secondly as a blueprint detailing the successive folds that turn a plain sheet of paper into some folded object.

The person who wants to design a paper glider using artificial evolution would then start with perhaps 30 sheets of paper, and write on each piece a random sequence of the symbols. Then she would take each piece of paper in turn, interpret the symbol string, the artificial DNA, as instructions to fold the paper, and see what shape results. The next step is to open a window high up in a building, and throw all 30 folded shapes out of the window.

She would then go outside and see how the different shapes have fallen to the ground below the window. Some may have fallen straight down, some may have accidentally been caught by some wind, some shapes may have possibly glided a metre or two. This is where Selection comes in, and the ones that have not traveled far are discarded while the one that went furthest are chosen to form the parents for the next generation. A new set of 30 sheets of paper is prepared, and strings of symbols, of artificial DNA, are copied onto them based on the surviving parents from the previous generation. This can be done in a variety of ways, any of which are likely to work.

The simplest option might be the asexual one, in which perhaps the best 50% of the previous generation each have 2 offspring, who inherit their single parent's DNA with some small probability of a mutation altering, deleting or adding a symbol. Alternatively, a form of sexual reproduction can be used, wherein the parents are brought together in pairs, and their offspring inherit some DNA from each parent, again with the possibility of further mutations. As long as the method chosen maintains the population of the next generation at the same size as the same as the initial generation, and obeys the rules for Heredity and Variation, then the stage is set for a further round of Selection on the new generation. Continuing this over many successive generations should result in increasingly successful paper gliders that fly further and further out of the window.

You can change the problem to that of designing real aircraft wings, or control systems for robots; and you can change the set of symbols to a new set specifying the curvatures and thicknesses of parts of a wing, or the type and connectivity of artificial neurons in an artificial neural net. Then the underlying methodology of Artificial Evolution will basically remain the same, except that the Selective process, the evaluation of the fitnesses of different members of the population, is likely to be far more expensive than throwing paper gliders out of the window.

When you change to a different problem, you have to create a new and appropriate method for interpreting strings of symbols, the artificial DNA, as potential solutions to the problem. For some problems it may be appropriate to use real-valued numbers as symbols in the DNA, in which case there is a potentially infinite range of values at such a locus on the genotype. In the work discussed from here on, however, it is explicitly assumed that, as in natural DNA, there is only a limited range of symbols, quite possibly limited to the binary range of 0 and 1.

3   The Microbial Genetic Algorithm

There are many varieties of Evolutionary Algorithms, many different ways to implement, for problem solving, the three main requirements of Heredity, Variation and Selection. I shall now describe one little known but effective method, that is so simple to implement that the core of the program can be reduced to a single line of code. I call it the Microbial Genetic Algorithm because it is loosely based on the way microbes can exchange genetic material, DNA, 'horizontally' between different living members of the population as an alternative to 'vertically' from one generation to the following one.

There are three particular tricks used here that are subtlely different from the basic algorithm described above in the paper gliders example. The first is the use of a 'Steady State' method rather than a 'Generational' method. Instead of accumulating a complete new generation of offspring, and then discarding the older generation and replacing it wholesale by the new, it is very reasonable to just generate a single new offspring at a time; then (in order to maintain the population size constant) choose one member of the population to die and be replaced by the new one. The Selection criterion will be satisfied by either biasing the choice of parent(s) for the new offspring towards the fitter members, or biasing the choice of which is to die towards the less fit. There are at least two advantages of the Steady State method over the generational method: it is usually much easier to implement, and it allows for efficient parallel implementations where it is not actually necessary to keep the evaluations of all members of the population in step with each other. Despite the fact that the generational method is usually the first to be discussed in the textbooks, these advantages mean that many serious users of evolutionary algorithms favour the Steady State method.

The second trick is to use a rank-based method of selection, and in particular tournament selection. The textbooks generally present 'fitness-proportionate' selection (where for instance if one member has twice the fitness of another member of the population it can expect twice as many offspring) as the main method used in GAs. This is probably for historical reasons, and because the formal analysis of GAs is mathematically easier when using this method. However, professional practitioners are far more likely to use a rank-based selection method, where the expected number of offspring of any member is based on (in the simplest case, linearly proportionate to) its ranking in the population. To give a simple example with a population of size 5, they can be ranked in order on the basis of their fitness and then allocated an expected number of offspring in this

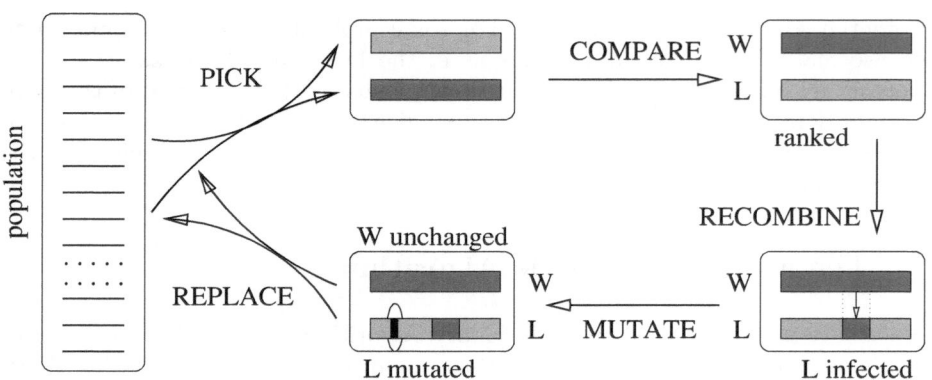

Fig. 1. A single tournament in the Microbial Genetic Algorithm.

ratio: 4/2 3/2 2/2 0/2. In this fashion the top-ranking member will have twice the expected number of offspring of the middle-ranking member, irrespective of whether it is 100 times fitter or only 1% fitter.

A cheap and cheerful method of implementing this type of rank-based selection, particularly appropriate for the Steady State case, is to pick out 2 members of the population at random and compare their fitnesses in a 'Tournament'. Then picking the winner to be a parent (or alternatively, picking the loser to be the individual that dies to make way for a new offspring) gives exactly the same expected selection bias as described in the previous paragraph. There are at least three advantages of this Tournament Selection method over the orthodox fitness proportionate selection method: it is usually much easier to implement, it avoids many scaling problems of the standard method, and it implements a form of elitism for free. Elitism in this context means that the currently fittest member of the population will always remain preserved unchanged.

Now we build on these two tricks by moving on to the third trick of the Microbial GA. It is perfectly acceptable to operate a GA by picking two members at random to be parents and generate a new offspring; and then pick a further two members at random, and using Tournament Selection choose the loser to die and be replaced by the new one. It may seem initially strange to have no bias towards choosing fitter members as parents, but the bias in choosing who is to die is what satisfies the criterion of Selection. The trick here is to combine all this into one operation.

So the Microbial method is to pick just two members of the population at random, who will be parents of the new offspring; and the least fit of the two parents is chosen as the one to die and be replaced. I have used so far the conventional language of 'parent', 'offspring' and 'die', but in fact this is equivalent to horizontal transmission of genetic material from the 'Winner' of the tournament to the 'Loser'. The Winner remains unchanged in the population, and the Loser receives copies of some genetic material (not necessarily restricted to 50%) from the Winner, with the opportunity for further mutations also.

The Microbial Genetic Algorithm is illustrated in the diagram, where the population of genotypes of 'artificial DNA' is represented by the set of lines on the left. Initially these will each be a random string of symbols, for instance a random binary string, and there will be some method for translating any such string into a trial solution for the design problem being tackled. This is where the human designer has to be creative in matching the genotype-to-phenotype translation to the requirements of the task. But then, provided that there is a suitable method for testing and scoring any such potential solution, giving it a 'fitness', all the rest of the work can be left to the algorithm. Two strings are picked out at random, and evaluated to see which is the Winner and which the Loser (W and L on the diagram). Then with some probability each locus (genotype position) of the Winner may be copied over the corresponding locus of the Loser, followed by a separate mutation process of changing at random some small proportion of the Loser loci. The two strings are re-inserted into the population — in fact the Winner is unchanged.

This Microbial GA obeys the 3 rules of Heredity, Variation and Selection, is effective, yet is so simple that it can be reduced to a single line of code. If we assume that, in C, the genotypes are in a binary array $gene[POP][LEN]$, and that the function $eval(i)$ returns the fitness of the ith member of the population, the one-liner goes something like this:

```
for (t=0;t<END;t++)
    for (W=(eval(a=POP*rand())>eval(b=POP*rand())?a:b),
        L=(W==a?b:a),i=0;i<LEN;i++)
            if ((r=rand())<REC+MUT)
                gene[L][i]=(r<REC ? gene[W][i] : gene[L][i]\^1);
```

4   Searching through Fitness Landscapes

Evolutionary algorithms, including the Microbial GA, can be thought of as search methods in a high-dimensional search space. Turning back to the paper glider folding example, if there are 8 possible folding instructions, and a succession of 25 folds, then there are 8^{25} possible versions of folding a glider. Only a tiny proportion of these will have any sort of flying ability, and an even smaller proportion will fly properly. If one considers all the 8^{25} designs as spread out over a landscape, with similar designs (differing by say just one fold) nearby to each other, then one can imagine the search process as searching across this landscape. Now treat the 'fitness' of each possible design as the 'height' of the corresponding position in this landscape, and we have a hilly fitness landscape where the peaks represent our goal. Typically the majority of this landscape will be foothills of negligible height, but it is reasonable to expect that the higher mountains form connected ranges that are the areas on which the search should be focused.

There are many possible strategies for searching such fitness landscapes, including Simulated Annealing, Hill-Climbing, Tabu Search. Since the search spaces are too big to search exhaustively, then all search methods involve sampling in turn successive points, checking their fitnesses and using this knowledge to guide the continuation of the search from what has been explored so far. The distinctive feature of evolutionary approaches such as genetic algorithms is the use of a *population* of search points, searching in parallel although not independently.

5   Conventional Genetic Algorithm Assumptions

If you read the Genetic Algorithm textbooks, you will find (explicitly or implicitly) a number of assumptions as to what makes the GA work effectively. One of the major worries is that of getting stuck on a local optimum in a fitness landscape; indeed this is the reason that most people are sceptical about simple Hill-Climbing methods.

It is generally assumed that GAs make a good effort to avoid getting trapped on such local optima through two properties. Firstly, by starting with an initial randomly spread population, there is more chance that the foothills to many different ranges will be encountered. The parallel population search will be eventually won by those climbing the mountain range that turns out to be the highest of those seen, and there is less chance of being trapped in one of the lower ranges.

Secondly, when there is recombination between different members of the population, this means that the searches are not truly independent. Even if two different members are in effect trapped on separate foothills (or local optima), then their offspring will, through recombination, occupy a new point on the fitness landscape somewhere that is in effect in-between these foothills. Hence such an offspring could escape from the local traps that each of its parents might be in.

From these ideas flow some further assumptions, widespread in the GA literature, that I shall argue are completely misleading. One major, and mistaken, worry is about 'premature convergence'. If you follow the above intuitions about how a single member of the population may get trapped in a local optimum, then you *need* a widely varied population to avoid this problem. Once all the variation in the population has disappeared over time, so that it is in effect multiple copies of the same individual, then it can get stuck on a local optimum however big the population of clones is. Unless new variation is injected into the population, then this genetic convergence will happen; and if it happens before the global optimum has been found, then this is the disaster of so-called 'premature convergence'. I shall give a different picture below.

The GA textbooks usually present a theorem derived by John Holland, the architect of GAs, called the *Schema Theorem*. This proves that under specific limited circumstances the 'useful parts' of genotypes in the population will grow exponentially as the GA produces the next generation from the current one. Although formally correct, it is usually misinterpreted as if this exponential growth continues unchecked over successive generations, whereas in fact it is only valid for a single generation; the calculations of fitnesses within the population have to be done afresh each time. So although the Schema Theorem is formally correct, pragmatically it is useless and irrelevant.

The Schema Theorem is associated with a commonly held dogma in the GA community, that *recombination*, the mixing and matching of various parts of the genotype from different parents to produce the offspring, is the driving force of evolutionary search. This feeds back to the worries about premature convergence spelled out above. There is an alternative viewpoint, however. Those who advocate the evolutionary methods of Evolutionary Programming (or EP) tend to emphasise the role of *mutation* rather than recombination, and for slightly different reasons so do I here.

6   SAGA: Species Adaptation Genetic Algorithms

In the natural world, of course, evolving populations are genetically highly converged. If this was not so, then the task of the Human Genome Project, assembling the genotype of a typical human being, would be pointless. The genetic differences between two human beings (or two members of any other species) are of course significant. They lie behind the subtle differences of human form and behaviour, of eye colour and temperament; they allow the possibility of DNA identification. But these differences are tiny compared to the similarities, to what makes an identifiable and coherent *species*.

Each and every species on this planet probably shares a common origin some 4 billion years ago. From this origin of life, variations have branched out with many such branches terminating, as species become extinct. But if we imagine following the historical trace of a currently-existing species, such as humans, we will find that for some 4 billion years there has been the phylogenetic pathway of a population changing from a single cell to the complex creatures we are today. At every point in this history, this population would have been genetically very converged, the genetic differences between individuals would have been minimal compared to their similarities.

In thought experiment at least, we could imagine this historical trace represented by a single individual from each generation, displaying our phylogenetic history. This history would be one of long-term change almost entirely through mutation; the interesting possible exceptions being when transfer of genetic material between species may occasionally bring together branches of the Tree of Life after they have previously bifurcated. So apart from such exceptions, all the accumulated design of a present day organism has come through the occasional lucky mutations that have been incrementally incorporated. Whatever the role of recombination might be in natural evolution — and the jury is still out on this question — it is mutation that is the driving force.

With this in mind, some ten years ago I started to develop a framework using similar ideas for long-term artificial evolution, for the incremental design methodology needed for such tasks as Evolutionary Robotics. This I call SAGA, or Species Adaptation Genetic Algorithms, as a genetically converged population, in effect a species, is involved. The early stages of SAGA [3,4] were based

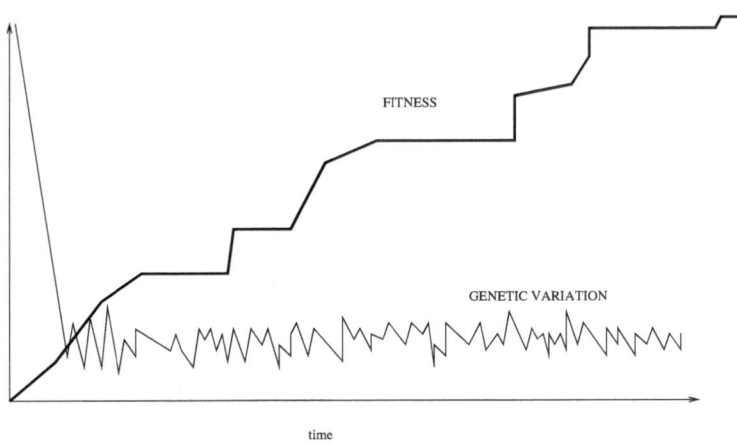

Fig. 2. Fitness typically continues to increase after the genetic variation has settled down.

on the realization that if long-term evolution meant that genotype lengths were initially relatively small (for encoding e.g. relatively simple robot control systems) and then slowly increased in length over generations to accommodate more complex designs as evolution progressed, then it was inevitable that the population would be genetically converged throughout. But later it came to be recognised that actually even with genotype lengths remaining constant, in practice an evolving population is genetically converged in any case. So it has turned out that SAGA ideas are far more widely applicable than their original domain.

7   Evolving a Genetically Converged Population

In artificial evolution, typically one starts with a randomly dispersed initial population, sampling widely across the search space. One useful measure of the genetic variation in a population is the average Hamming distance between two members; the average number of places on the genotype where there genetic material is different. For a random population of binary genotypes, any two will differ in about half the positions, but as successive rounds of selection discard the less fit and concentrate the population around the fitter ones, this variation will dramatically decrease. If one plots this genetic convergence in a practical example of artificial evolution, then one sees just how surprisingly fast this occurs; even with a large population, most of the variation will have disappeared within 10 generations or so.

Yet if one also plots the fitness of the population over time, one can see that the fitness continues to increase after the genetic variation has been reduced to its (noisy) minimum range of values.

In the absence of any mutation, selection will concentrate the population at the current best. The smallest amount of mutation will hill-climb this current best to a local optimum. As mutation rates increase, the population will spread out around this local optimum, searching the neighbourhood; but if mutation rates become too high then the population will disperse completely, losing the hill-top, and the search will become random. If a balance is achieved between selective forces and those of mutation (as modified by recombination), then some elements of the population can crawl down the hill far enough to reach a ridge of high selective values. As discussed in [2], this results under selection in a significant proportion of the population working their way along this ridge, and making possible the reaching of outliers further in Hamming-distance in that particular direction from the current fittest. The term 'ridge' is used here to fit in with intuitive notions of fitness landscapes; in fact in high-dimensional search spaces such ridges may form complex neutral networks, percolating long distances through genotype space.

If any such outliers reach a second hill that climbs away from the ridge, then parts of the population can climb this hill. Depending on the difference in fitness and the spread of the population, it will either move *en masse* to the new hill as a better local optimum, or share itself across both of them.

So in a SAGA setup of evolution of a converged species, we want to encourage through the genetic operators such hill-crawling down towards ridges to new hills, subject to the constraint that we do not want to lose track of the current hill. Eigen and co-workers use the concept of a quasi-species to refer to a similar genetically converged population in the study of early RNA evolution. To quote from [2]:

> In conventional natural selection theory, advantageous mutations drove the evolutionary process. The neutral theory introduced selectively neutral mutants, in addition to the advantageous ones, which contribute to evolution through random drift. The concept of quasi-species shows that much weight is attributed to those slightly deleterious mutants that are situated along high ridges in the value landscape. They guide populations toward the peaks of high selective values.

8   SAGA and Mutation Rates

Although progress of a species through a fitness landscape is not discussed in the standard GA literature, in theoretical biology there is relevant work in the related field of molecular quasi-species [1,2]. In particular, analysis of 'the error catastrophe' shows that, subject to certain conditions, there is a maximum rate of mutation that allows a quasi-species of molecules to stay localised around its current optimum. This critical maximum rate balances selective forces tending to increase numbers of the fittest members of the population against the forces of mutation that tend, more often than not, to drag offspring down in fitness away from any local optimum. But a zero mutation rate allows for no further local search beyond the current species, and other things being equal increased mutation rates will increase the rate of evolution. Hence if mutation rates can be adjusted, it would be a good idea to use a rate close to but less than any critical rate that causes the species to fall apart. A further possibility, in the spirit of simulated annealing, is to temporarily allow the rate to go *slightly* above the critical rate — to allow exploration — and then cut it back again to consolidate any gains thus made.

For an infinite asexual population, it can be shown (e.g. in [1]) that these forces just balance for a per-genotype mutation rate m equal to the logarithm of σ; where σ is the *superiority* parameter of the fittest member of the population — the factor by which selection of this sequence exceeds the average selection of the rest of the population. Recombination makes some degree of difference [5], but the end result stays in the same general area. The rule of thumb is that if the selection pressure used is that associated, for instance, with the Microbial GA using tournaments of size 2, the optimal mutation rate is in the region of one mutation per genotype, after taking account of any junk or neutral mutations (see below). In other words, the rate should be set so as to expect around one fitness-altering mutation in the whole genotype; if, for example, 50% of the genotype is redundant or neutral, such that mutations in those regions make no difference, then a rate of 2 mutations per complete genotype gives an expected one mutation in the non-redundant part.

When applying such mutation rates in a GA, it is essential that the probability of mutation is applied independently at each locus on the genotype. This gives a binomial distribution (approximating a Poisson distribution for long genotypes) for the number of mutations per string, so that genotypes with an expected m mutations have this as the average value with a wide variance (including the possibility of zero mutations).

9   Neutral Networks and Drift

Mutations in a genotype encoding a fit phenotype are often deleterious, and occasionally advantageous. There is a third possibility, that a mutation is neutral and leaves the fitness unchanged.

Neutral mutations can in turn be subdivided into two kinds, with a rather grey area between them. They can be in parts of 'junk DNA', such that the decoding of the genotype ignores the values in that part. In this case it is only the functional part of the genotype, the part that is capable of causing some difference in the fitness, that counts towards effective genotype length when deciding upon mutation rates. For example, if a genotype of length 1000 is 90% junk, then a mutation rate set at the rate of one per effective genotype length should be implemented at the rate of 1/100 per locus, rather than 1/1000. It is often difficult to estimate what proportion of a genotype is junk, however, as this shades into the second class of neutral mutation.

This second type of mutation may leave the phenotype unchanged, yet open the possibility of a further mutation making some difference. At its simplest level, a binary genotype with two loci, whose fitness is given by the logical **AND** of the alleles at each locus, retains a fitness of 0 during mutation from **00** to **01**; yet this opens up the possibility of a further single point mutation reaching **11**, with a fitness of 1 which was not achievable from the starting point. Such neutral mutations can in a high-dimensional space allow extended neutral paths that can percolate through vast areas of sequence space. Neutral drift of a population through such pathways means that it is much more difficult than one might think to get stuck on a local optimum. In addition, the percolation of such paths through sequence space tends to mean that it does not matter too much where in sequence space a converged population starts; under many circumstances it is possible to reach all possible fit regions from most starting points.

The SAGA selection and mutation rates encourage just such exploration through neutral drift in sequence space.

10 Recombination

With a genetically converged population, sections of genotype that are swapped in recombination are likely to be fairly similar. With species evolution recombination does not have the prime significance it has in standard GAs — asexual evolution is indeed feasible — but nevertheless it is a useful genetic operator.

There are two roles recombination has which are opposite sides of the same coin. On the one hand, it allows two fortunate mutations that happen to have occurred independently in two different lineages within the population to be combined into one which has both; something not possible with asexual reproduction. On the other hand, it allows parents with a detrimental mutation to produce an offspring which does not have it; also impossible asexually, in the absence of highly improbable back-mutations. This latter effect in general allows higher mutation rates to be used with recombination than were suggested above for asexual populations, thus promoting exploration without risking loss of a currently achieved local optimum.

Recombination is particularly powerful when combined with a distributed GA. Here each member of the population is allocated a different position in some notional geographical space, often a two-dimensional toroidal grid Recombination between individuals is only allowed for pairs within a certain distance of each other on this grid, which thus comprises a number of overlapping neighbourhoods. This combines the virtues of small and large populations; small interrelated local populations allows through random drift more extended search through genotype space, but the overlapping nature of such localities means that any improvement found percolates through the whole population.

11 But Does It Work?

The SAGA approach to artificial evolution assumes that evolution can incrementally improve the fitness of a population over the long term, despite the population being genetically converged. This requires the existence of ridges or neutral networks, to avoid getting trapped on local optima. Now clearly many fitness landscapes do not have these useful properties of neutral networks, of escape routes; indeed, almost all the benchmark problems and test suites in the GA literature do not have such useful properties. So why should one expect a difficult practical problem to have such neutral networks?

Here we should start by appealing to some mathematical intuitions, before progressing to look at an actual practical example. Suppose we have a problem encoded with binary genotypes of length 1000, so the genotype search space has 2^{1000} points and is impractically large to search exhaustively. If the problem is to find appropriate robot behaviours, and a genotype specifies a robot control system, then typically there will be *very* many different genotypes that will produce the same phenotype (or robot behaviour). For the sake of an example, let us suppose that the genotype is 50% redundant, implying that there are 2^{500} different phenotypes; and each of these can be specified, on average, by 2^{500} different genotypes. One can think of this as a 500-dimensional phenotype space mapped into a 1000-dimensional genotype space.

If there is some degree of correlation, some tendency for similar phenotypes to map into similar genotypes, then one can expect the genotypes corresponding to some specific phenotype P to be partially correlated, and indeed largely connected in genotype space; in fact to form a neutral network. And it is entirely reasonable to expect some form of such correlation in any practical problem, because otherwise such problems would be completely intractable.

These are the intuitions, but in at least one example this has been tested in practice. Adrian Thompson has pioneered intrinsic Hardware Evolution at Sussex [6], using artificial evolution to design electronic circuits for pattern recognition tasks on reconfigurable silicon chips. The evolutionary method used was based on SAGA principles outlined above. In one experiment he deliberately set out to test the hypothesis that one can reasonably expect there to be pathways through genotype space to a (near-) perfect solution, that do not get trapped in local optima. With this in mind, he used a population size one in a form of Hill Climbing.

A genotype specified the current population, and we can simplify the description somewhat to say that this was effectively a binary genotype of length 1900, in other words a genotype search space of 2^{1900}. At each step in the search process, a minimal mutation was applied to the genotype, and the new fitness compared to the previous one. If the mutation resulted in a decreased fitness, that step was abandoned; but if fitness either increased *or* remained the same, the step to the new mutant was taken. This means that the eventual pathway seen through genotype space consisted of only upward or horizontal steps in the fitness landscape; in practice far more horizontal ones than upward. In fact the fitness graph took the typical form of punctuated equilibria.

The mutational steps were not actually single mutations. On SAGA principles, the mutation rate should be set so as to generate an expected one fitness-altering mutation per genotype. It was already known, from previous experiments, that around 2/3 of the genotype was redundant, so that a mutation rate

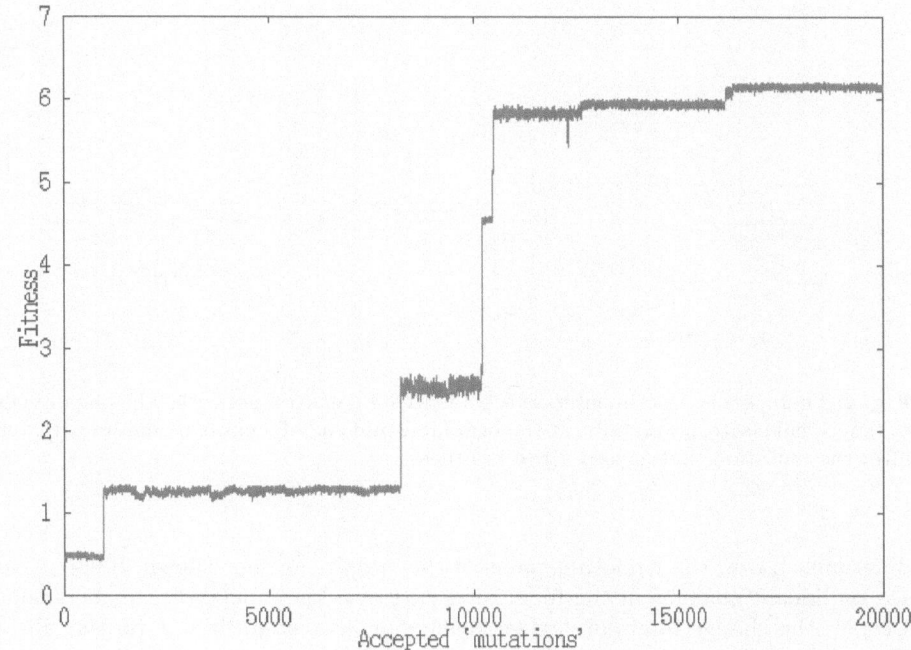

Fig. 3. The graph of fitness rising over successive mutational steps, with plateaus or 'equilibria' punctuated by rises in fitness.

of around 3 mutations per genotype could be expected to produce on average one fitness-altering one. A further complexity that we need not pursue further here was that fitness measurements took place on the real silicon chip, with the inevitable noise associated with real physical processes; this meant that care was needed in taking re-evaluations of the fitness when it was felt that a previous one may have been, through noise, misleading.

As can be seen from the fitness graph, there was indeed a connected pathway from the low-fitness starting point to the final good solution. Much of the time was spent on the level sections where fitness basically remained constant (within the limits of noise and measurement). Now possibly this time was wasted, merely mutating redundant parts of the genotype until a lucky break was hit upon. Thompson set out to test whether this was the case by looking at what happened as the search traveled along a specific plateau.

This plateau started at generation 13000, and continued for over 3000 successive 'accepted mutations' until between generation 16144 and 16145 a new fitness-increasing mutation was found. Along the plateau there was clearly a lot of genetic change, and it was possible to investigate the associated phenotypes, or useful functional parts of the genetically specified electronic circuits. In the

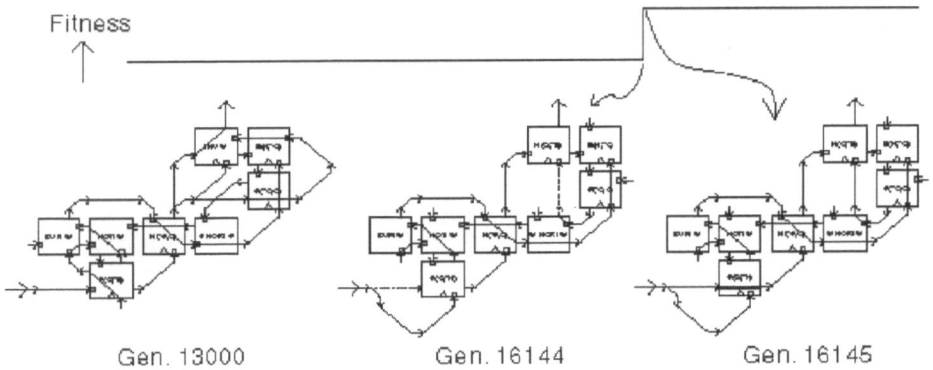

Fitness

Gen. 13000   Gen. 16144   Gen. 16145

Fig. 4. Thompson's experiment to seek evidence for a neutral network. The phenotypes, or functional electronic circuits, at the beginning and end of a plateau; and immediately after the mutation that caused a rise in fitness.

diagrams shown, the irrelevant parts of the circuits are not shown, but one can see significant changes in the functional part over these 3000 or so generations. Despite the change in circuits, their behaviour, as tested directly on the silicon chip faced with a signal recognition task in real time, was unaltered.

The plateau ended with a single lucky mutation that increased fitness, so the interesting question was posed: would that specific mutation have been as lucky if it had occurred earlier on in that plateau? The mutation was applied to generation 13000, and it was observed that fitness fell significantly as a result, rather than increased. Further experiments showed that there was indeed no possible lucky single mutation from generation 13000, and from this we can deduce that indeed the drift along the neutral network was indeed beneficial. By analogy, if we wish to reach the top of a multi-storey building, then walking along a level corridor may not immediately gain us height, but will still be beneficial if it leads us to a distant upwards stairway.

This experiment gives support to the conjecture that in some real problems with very high-dimensional search spaces, using a genetic code with a discrete alphabet of symbols, there may indeed be useful neutral networks that assist in avoiding getting trapped in local optima.

12   Conclusions

I have given a very basic sketch of the underlying principles of artificial evolution, and illustrated how simple this can be by demonstrating the very minimalist Microbial Genetic Algorithm. I have argued for a rather different perspective on GAs than that conventionally presented. One should expect evolution to proceed even when a population is to a great extent genetically converged. SAGA principles suggest that mutation rates should be tuned to give an optimum amount of genetic convergence, arising form a balance between the inward forces of Selection, and the outward forces of mutation that add Variation. These principles have a wide applicability.

There are conventional worries about genetic convergence, often phrased in terms of 'premature convergence'. There are good reasons for believing that these worries are ill-founded in some very high-dimensional search spaces. Neutral Networks offer potential escape routes, and an example shows how this can work in practice.

Acknowledgments. I thank colleagues in COGS at the University of Sussex for fruitful discussions and collaborations on issues related to this paper.

References

1. Eigen, M. New concepts for dealing with the evolution of nucleic acids. In *Cold Spring Harbor Symposium on Quantitative Biology*, vol. LII, 1987.
2. Eigen, M., McCaskill J., and Schuster, P. Molecular quasi-species. *Journal of Physical Chemistry*, 92:6881-6891, 1988.
3. Harvey, I. Species Adaptation Genetic Algorithms: A basis for a continuing SAGA In *Toward a Practice of Autonomous Systems: Proceedings of the First European Conference on Artificial Life*, F.J. Varela and P. Bourgine (eds.). MIT Press/Bradford Books, Cambridge, MA, 1992, pp. 346-354.
4. Harvey, I. Evolutionary robotics and SAGA: the case for hill crawling and tournament selection. In Langton, C. (ed.), *Artificial Life III, Santa Fe Institute Studies in the Sciences of Complexity, Proc. Vol. XVI*, pp. 299–326. Addison Wesley.
5. Ochoa, G., I. Harvey, I., and H. Buxton, H. Error Thresholds and their Relation to Optimal Mutation Rates. In *Proceedings of the Fifth European Conference on Artificial Life (ECAL99)*, D. Floreano, J-D. Nicoud, F. Mondada (eds). Springer-Verlag, 1999.
6. Thompson, A. Hardware Evolution: Automatic design of electronic circuits in reconfigurable hardware by artificial evolution. *Distinguished dissertation series*, Springer-Verlag, 1998.

Using Biological Inspiration to Build Artificial Life That Locomotes

Robert J. Full

Department of Integrative Biology
University of California at Berkeley
Berkeley, CA, 94720, USA
rjfull@socrates.berkeley.edu

Abstract. Nature's general principles can provide biological inspiration for robotic designs. Biological inspiration in the form of genetic programming and algorithms has already shown utility for automated design. However, reliance on evolutionary processes mimicking nature will not necessarily result in designs better than what human engineers can do. Biological evolution is more like a tinkerer than an engineer. Natural selection is constrained to work with pre-existing materials inherited from an ancestor. Engineers can start from scratch and select optimal raw materials and tools for the task desired. Nature provides useful hints of what is possible and design ideas that may have escaped our consideration. The discovery of general biological design principles requires a collapse of dimensions in complex systems. Reducing redundancies by seeking synergies yields simple, general principles that can provide inspiration. Even if we had all the general biological principles, we don't have the technology to use them effectively. Information handling has changed dramatically, but until recently the final effectors (metal beams and electric motors) have not. Nature will become an increasingly more useful teacher as human technology takes on more of the characteristics of nature. The design of artificial life will require unprecedented interdisciplinary integration.

1   Introduction

Evolutionary robotics appears to have evolved. The commonly held tenets now point toward the attainment of self-organized, autonomous, situated, interactive machines [7]. The field's collective focus seems to have progressed from a branch of artificial intelligence involving machine learning that was virtual to an emphasis on situated and embodied robotics [4]. Yet, within the field of evolutionary robotics at least three more distinct goals have been articulated [7].

Artificial Life. The artificial life community desires to create life-like creatures and life-as-it-could-be. Those attempting to create artificial life value full autonomy, self-sufficiency and self-containment. Many terms and concepts have been borrowed in part from the field of evolutionary biology in an effort to reveal

T. Gomi (Ed.): ER 2001, LNCS 2217, pp. 110–120, 2001.

possible patterns of life. Obviously, this effort impacts algorithm development in fields of learning, adaptive control and optimization. By bringing novel ideas and methodologies, perhaps integration with evolutionary biology will lead to novel hypotheses for understanding the evolution and behavior of real life.

Assisting biologists with physical models. For centuries, biologists have borrowed ideas from physics, mathematics and engineering [24]. Truly remarkable discoveries have been made about how organisms work by developing testable hypotheses from knowledge in the physical sciences. Advances in the fields of physiology and biomechanics provide the most striking examples. In particular, the use of physical models to elucidate complex phenomenon still plays a major role despite our increasing capacity for accurate simulation. For example, the paradox of insect flight was resolved, not by solving three-dimensional Navier-Stokes equations, but by flapping scaled-model wings in a vat of syrup [6]. Our own research on legged locomotion has benefitted directly from the construction of several robots by providing us new hypotheses of control, stability and adhesion [8]. The use of physical models from evolutionary robotics directed toward the evolution and behavior of organisms promises to deliver novel hypotheses that may explain complex biological phenomenon. Attainment of this goal will continue to foster the exchange of ideas that will benefit several communities.

Attaining automated engineering. Artificial evolution can assist in automatically developing algorithms and machines that display complex, life-like capabilities that would be otherwise difficult to program [7]. Evolutionary techniques have been successfully applied in diverse areas such as network management, insurance, elevator operation, and circuit design. A more ambitious goal strives for "full autonomy . not only at the level of power and behavior, but also at the levels of design and fabrication"[18]. The justifications for using artificial evolution in engineering are varied. One extreme view claims that human engineers have failed and will continue to fail. "Robots are still laboriously designed and constructed by teams of human engineers, usually at considerable expense. Few robots are available because these costs must be absorbed through mass production, which is justified only for toys, weapons and industrial systems such as automatic teller machines [18]." The implication being that an artificial evolution approach would necessarily yield better results.

Engineers may not have created robots that operate as effectively as we have imagined or robots that are an economic success. However, the reliance on evolutionary processes mimicking nature will not necessarily do better. Biological evolution operates more on sufficiency rather than optimality. Engineers have distinct advantages over evolutionary processes. Secondly, an approach where biology inspires engineering holds greater promise. Biological inspiration involves the transfer of biological principles to engineers who are capable of capitalizing on them. Although nature is complex, general principles, rules and mathematical models can be extracted and novel designs characterized. Biological inspiration should include concepts from evolution, adaptation and learning. Thirdly, one of the reasons we may be dissatisfied with present day robots is that until now nature could not be a very good teacher because human technology differed so from natural technology.

2   Evolutionary Tinkering vs. the Human Engineer

Biological inspiration in the form of methodologies such as genetic programming and genetic algorithms has already shown utility for automated design. However, reliance on evolutionary processes mimicking nature too closely will not necessarily result in designs better than what human engineers can do. It is important to be reminded that biological evolution works on the "just good enough" principle. Organisms are not optimally designed and natural selection is not engineering [14]. Engineers often have final goals, whereas biological evolution does not. Organisms must do a multitude of tasks, whereas in engineering executing far fewer tasks will do. As a result, "trade-offs" are the rule, severe constraints are pervasive and global optimality rare in biological systems.

2.1   Constraints and Biological Evolution

Biological evolution has brought us amazingly functional and adaptive designs. However, we must not forget that about five hundred million species have gone extinct and only a few million remain. Biological evolution works more as a tinkerer than an engineer [16]. The tinkerer never really knows what they will produce and uses everything at their disposal to make something workable. Organisms carry with them the baggage of their history. Therefore, they must co-opt the parts they have for new functions. Part of an ear is built from jaw bones and wings from legs. Organisms are not an optimal product of engineering, but "a patchwork of odd sets pieced together when and where opportunities arose" [16]. Natural selection is constrained to work with the pre-existing materials inherited from an ancestor. Dolphins have not re-evolved gills and no titanium has been found in tortoise shells [14]. Engineers can start from scratch and select the optimal raw materials and tools for the task desired, natural selection can not.

Organisms are not optimally adapted for the environment in which they reside. Biological evolution can't keep pace with the changing environments because not all phenotypic variation is heritable and if selection were too strong it could easily produce extinction. Natural selection can't anticipate major changes in environments. Behavior can evolve more quickly than morphology and physiology leading to mismatches. Engineers can optimize for one or a few environments and choose to add appropriate safety factors as dictated by previous experience.

Finally, most organisms grow, but must continue to function. As a result development can constrain evolution of the final product — the adult. Engineers are not so constrained and fortunately are not required to make fully function miniature versions of their final designs.

2.2   Nature's Role

Nature provides useful hints of what is possible and design ideas that may have escaped our consideration [24]. Given the unique process of biological evolution and its associated constraints, identifying, quantifying and communicating these design ideas is a challenge. Here is where the integrative biologist can contribute most to the inspiration transferred to the engineer. Biologists offering advice need not only understand principles of structure and function, but use their knowledge of phylogenetic analysis, behavior and ecology to extract potentially valuable design ideas. Design ideas motivated from nature should include those involving the processes of development, evolution and learning. However, engineers should not blindly copy these design ideas. In many cases, engineers have developed approaches, tools, devices and materials far superior to those in nature.

3   Biological Inspiration — An Example from Legged Locomotion

Nature's general principles can provide biological inspiration for robotic designs. The challenge is to discover them.

3.1   Curse of Dimensionality

The discovery of general biological principles often requires a collapse of dimensions because of the complexity of organisms. Behavior results from complex, high dimensional, nonlinear, dynamically coupled interactions of an organism with its environment. Given an organism's capacity for a multitude of behaviors together with the remnants of their history, we should not be surprised that animals show a remarkable degree of apparent redundancy when a single behavior such as locomotion is examined. Animals show kinematic redundancy for locomotion, because they have far more joint degrees of freedom than their three body positions and three body orientations. Animals show actuator redundancy for locomotion, because they often have at least twice as many muscles as joint degrees of freedom. Animals show neuronal redundancy for locomotion, because they have more participating interneurons than required to generate observed motor neuron signals. Reducing redundancies by a collapse of dimensions aided by seeking synergies and symmetries can yield simple, general principles. For instance, animals that differ in leg number, body form and skeletal type show the same dynamics of the center of mass [9]. All rapidly moving legged animals bounce like people on pogo sticks. Force patterns produced by six- legged insects are the same as those produced by trotting eight-legged crabs, four-legged dogs and running humans. Each animal has two sets of virtual legs that alternate. One leg of a humans works like two legs of a trotting dog, three legs of an insect and four legs of a crab.

3.2   Templates and Anchors

Fortunately, simple models we call *templates* can be made to resolve the redundancy observed [10]. A template is the simplest model with least number of
variables and parameters that exhibits a targeted behavior (Fig. 1).

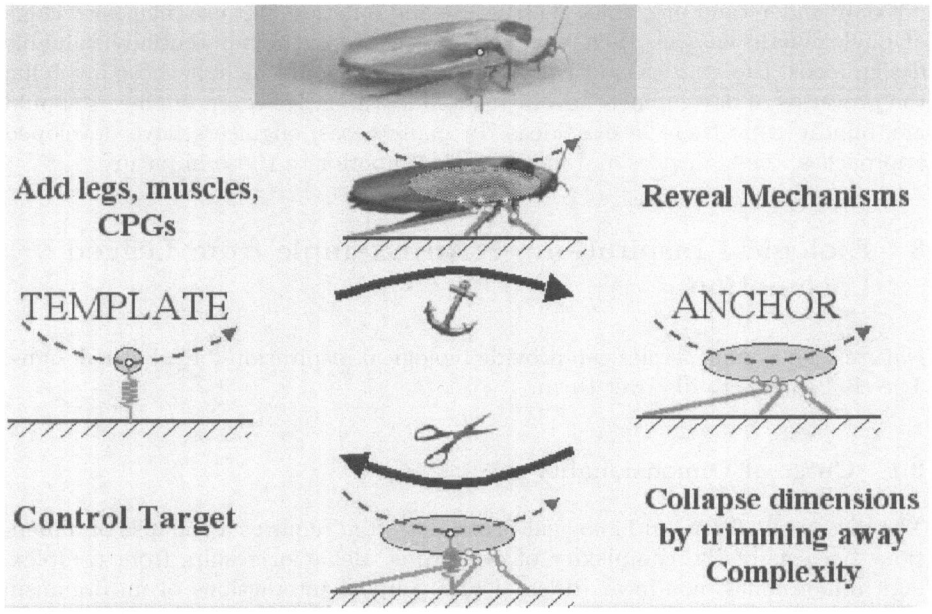

Fig. 1. Templates and anchors. The creation of templates and anchors for biological
systems generates testable hypotheses that allow us to address the complexity observed
in nature. This approach can be used to provide biological inspiration to engineers.

Templates suggest control strategies that can be tested against empirical
data. Yet, templates must be grounded in more detailed morphological and phy-
siological models to ask specific questions about multiple legs, the joint torques that actuate them, the recruitment of muscles that produce those torques,
and the neural networks that activate the ensemble. We term these more elaborate models *anchors*. They introduce representations of specific biological details
whose mechanism of coordination is of interest. Since mechanisms require controls, anchors incorporate specific hypotheses concerning the manner in which
"unnecessary" motion or energy from legs, joints and muscles are "trimmed
away" leaving behind the behavior of the body in the low degree of freedom
template.

3.3   Evolution of Legged Robots

Ariel (Fig. 2 left) is an example of a biologically inspired robot (built by iRobot) benefitting from basic research on animals (conducted at UC Berkeley). This sprawled posture, hexapod is the only mobility platform that can move on land and maneuver underwater in the surf zone much like true amphibians - crabs [15]. To inspire the robot's leg design, we collected data on the morphology and kinematics of locomoting crabs [19]. We discovered joint synergies that reduced the degrees of freedom for each leg from nine to two, greatly simplifying actuation and control. We also found that crabs vary their stance width to reduce overturning moments when faced with a more variable environment. The use of a variable stance width in Ariel increased stability as well as adding the capability of obstacle clearance. Ariel also benefitted from behavioral observation of crabs on sand in the surf zone. To station keep, crabs oscillate their legs, dig into the sand and then generate a lateral gripping force. Ariel can use this behavior, but when its body surface orientation is unimportant it uses a decidedly, non-biological approach. If flipped over, Ariel simply inverts its legs and continues on.

Fig. 2. Biologically inspired robots. *Left.* Ariel. The first amphibious, legged robot capable of maneuvering in the surf zone. Built by iRobot in collaboration with UC Berkeley. *Right.* RHex, The robot hexapod capable of negotiating rough terrain using only six degrees of freedom, feed forward control and no external sensing of the environment. Built by University of Michigan and McGill University in collaboration with UC Berkeley.

Running insects operate as spring-mass templates [12,13] and use a largely passive, dynamic, self-stabilizing mechanical system to rapidly maneuver over rough terrain [17,22,23]. We have shown that the high dimensional space in which these agile, many-legged animals operate can be collapsed down to a few degrees of freedom, thereby simplifying control. Our findings inspired engineers to design and construct a dynamic hexapod, named RHex (Fig. 2 right), with only a six degree of freedom body forced by six passive, springy legs, each driven by one geared DC servo [1,21]. RHex can maneuver over a forest floor, traverse sand dunes, negotiate rocks, climb over pipes and up stairs and even swim, all without feedback from the external environment.

4 Nature as a Teacher

If general principles can be extracted from nature, then why is there dissatisfaction with biologically inspired robots? As stated previously, a justification for primary use of artificial evolution for design and manufacturing points to the laborious efforts of engineers and the resulting financial burden. An alternative explanation lies in the difference in the technologies used by humans versus those observed in nature.

Nature will become an increasingly more useful teacher as human technology takes on more of the characteristics of nature [24]. Even if we had all the general biological principles, we don't have the technology to use them effectively. Information handling has changed dramatically, but until recently the final effectors (metal beams and electric motors) and sensors have not.

4.1 Human vs. Natural Technology

Traditionally, human technologies have been large, flat, right-angled, stiff, and rotating, with few actuators and sensors, whereas nature is small, curved, compliant using appendages with multiple actuators and sensors (Table 1, [24]). Human technology is changing with the greater use of nonmetallic, more flexible materials and increased miniaturization. Revolutionary new technologies in materials and manufacturing promises to lead to more life-like, mobile robots in the future when inspired by nature.

Table 1. Human vs. Natural Technologies

Human Technologies	Natural Technologies
Large	Small
Flat, Right-angled	Curved
Stiff	Bend, twist
Rotating devices	Appendages
Few sensors and actuators	Many sensors and actuators

4.2 Promise of New Materials and Manufacturing Techniques

Robots lack the robustness and performance of animals when operating in unstructured environments. However, even biologically inspired robot designs are compromised by the fragility and complexity that result from using traditional engineering materials and manufacturing methods. Clearly, designs must be combined with physical structures that mimic the way biological structures are composed, with embedded actuators and sensors and spatially-varied passive properties. A new layered-manufacturing technology called Shape Deposition Manufacturing (SDM) makes this possible [3]. SDM's unique capabilities have resulted in a family of hexapedal robots whose fast (over 4 body-lengths per second) and robust (traversal over hip-height obstacles) performance begins to compare to that seen in nature (Fig. 3).

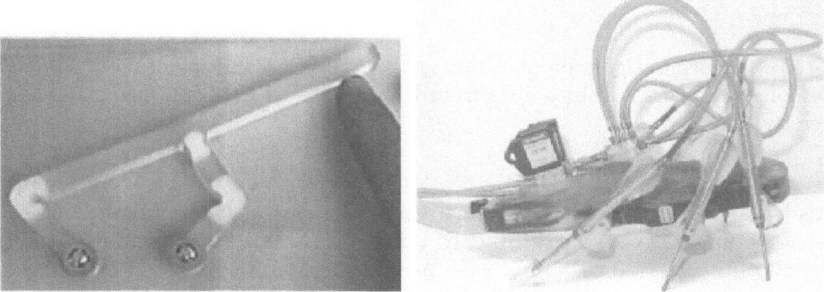

Fig. 3. Shape deposition manufacturing (SDM). Left. SDM compliant leg. Right. Sprawl. An SDM robot hexapod capable of negotiating rough terrain by taking advantage of the passive dynamic properties of its legs. No external sensing of the environment is required. Built by Stanford University in collaboration with UC Berkeley.

One major difference between animals and present day robots is found in the structure and performance of actuators. Animals use muscle, whereas most robots use motors. Muscle is a light-weight, multi-functional material. Muscles can function as motors, springs, struts and shock absorbers [5]. In collaboration with SRI International, we are currently measuring the muscle- like properties of electroactive polymer (EAP) actuators [11]. In these dielectric elastomers strain is induced through Maxwell stresses caused by the application of an electric field [20]. We have examined EAP actuators in the very same experimental apparatus in which we test natural muscle. We have discovered that acrylic EAP actuators produce stresses, strains, work and power outputs that fall within the capabilities of natural muscles.

Another extraordinary manufactured material that prevents almost any surface from being an obstacle to locomotion can be found on the feet of geckos. Each foot contains over 1 million tiny hairs arranged in rows [2]. Each hair can have as many as one thousand 0.2 micron flattened tips called spatulae. The nearly two billion spatulae allow such close contact with the surface that the adhesion force results from van der Waals interactions among molecules. Efforts are underway to use this inspiration to make the first self-cleaning, dry adhesive.

Research on gecko foot hairs did not originate with an effort to study adhesion. Instead, we were asked by a robotics company, iRobot, if we could provide biological inspiration toward the development of a climbing robot. We selected

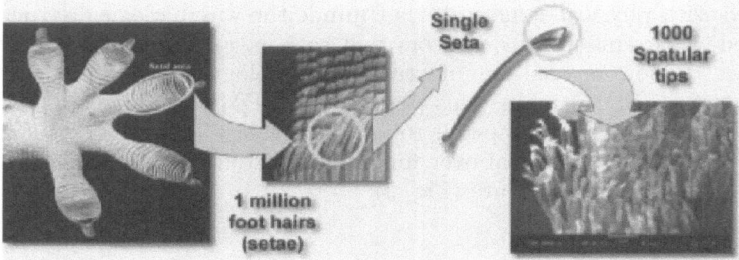

Fig. 4. Self-clearing, dry adhesive from gecko foot. Research a result of a collaboration among engineers and biologists from UC Berkeley, Lewis and Clark and Stanford University.

geckos because of their extraordinary ability to run up vertical surfaces at high speed. In doing so, we contributed to the evolution of an autonomous climbing robot, the first version of which was named the Mecho-gecko (Fig. 5). The success of the first clade of climbing robots was not due to the direct copying of the complex hair morphology, but was instead inspired by the observation that some geckos uncurl their toes during attachment and peel them away from the surface during detachment. By running geckos over a fancy scale (three dimensional force platform) imbedded in wall, we showed that large attachment and detachment forces were completely absent. By using this peeling strategy with a pressure sensitive adhesive, iRobot engineered two autonomous climbing robots. The next step is to evolve a robot with enhanced maneuverability for climbing. These robots will likely take advantage of legs and toes. In the future, we hope to evolve a climbing robot that will use the capacity offered by self-cleaning dry adhesives.

5 Age of Integration

Evolutionary robotics will continue to evolve. Biological inspiration can lead the way toward artificial life capable of extraordinary performance. Genetic programming, genetic algorithms and artificial evolution will continue to play a key role in the inspiration. Human discovery beyond artificial evolution will as well. Hopefully, we can direct artificial evolution along predominately productive lineages, and avoid the pitfalls so common in the history of natural life. Biologists working with engineers and mathematicians are discovering the general principles of nature from the level of molecules to behavior at an ever-increasing pace. Now more than ever before, nature can instruct us on how to best use new materials and manufacturing processes discovered by engineers, because these human technologies have more of the characteristics of actual life. This effort will require unprecedented integration among disciplines that include biology, psychology, engineering, physics, chemistry, computer science and mathematics. Fortunately, the age of integration is here.

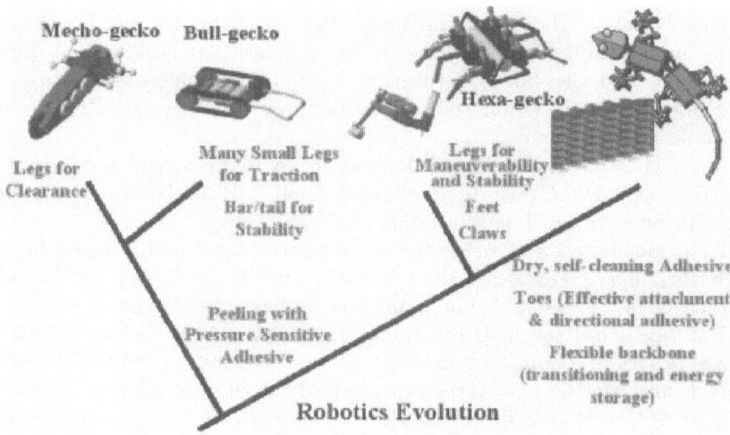

Fig. 5. Evolution of climbing robots. The evolution of the Mecho-gecko and the Bull-gecko included the innovation of using peeling with a pressure sensitive adhesive. The Hexa-gecko and Dry adhesive-gecko will be designed to use legs and toes. The Dry adhesive-gecko awaits the manufacture of the self-clearing dry adhesive hairs. Built or designed by iRobot in collaboration with UC Berkeley.

Acknowledgements. Supported by ONR MURI N00014-98-1-0669, DARPA/ONR N00014-98-1-0747 and DARPA/ONR N00014-98-C-0183 grants to RJF. Thanks to Kellar Autumn for reading the manuscript.

References

1. Altendorfer, R., Saranli, U., Komsuoglu, H., Koditschek, Brown Jr., B.H., Buehler, M., Moore, N., McMordie, D. and Full, R.J.: Evidence for Spring Loaded Inverted Pendulum Running in a Hexapod Robot. *Proceedings of the International Symposium on Experimental Robotics*, Honolulu, HI, (2000)

2. Autumn, K., Liang, Y., Hsieh, T., Zesch, W., Chan, W.-P., Kenny T., Fearing, R., and Full, R.J.: Adhesive force of a single gecko foot-hair. *Nature* **405** (2000) 681–685

3. Bailey, S.A., Cham, J.G., Cutkosky, M.R., and Full, R.J.: Biomimetic Robotic Mechanisms via Shape Deposition Manufacturing. In *Robotics Research: the Ninth International Symposium.* (eds. J. Hollerbach and D. Koditschek), Springer-Verlag, London (2000) 403–410.

4. Brooks, R.A., Breazeal, C., Irie, R., Kemp, C.C., Marjanovic, M., Scassellati, B. and Williamson, M.M.: Alternative essences of intelligence. Tenth Conference on Innovative Applications of Artificial Intelligence, *Proceedings of the Fifteenth National Conference on Artificial Intelligence*: AAAI Press/MIT Press, (1998) 961–968

5. Dickinson, M.H., Farley, C.T., Full, R.J., Koehl, M.A. R., Kram R. and Lehman, S.: How animals move: An integrative view. *Science* **288** (2000) 100–106

120   R.J. Full

6. Dickinson, Michael H., Lehmann, Fritz-Olaf and Sane, Sanjay P.: Wing rotation and the aerodynamic basis of insect flight. *Science* **284** (1999) 1954–1960
7. Floreano, D. and Urzelai. J.: Evolutionary robots: the next generation. Evolutionary Robotics. In T. Gomi (ed.), *Evolutionary Robotics III*, Ontario (Canada): AAI Books. (2000)
8. Full, R.J.: Biological inspiration: Lessons from many-legged locomotors. In: *Robotics Research 9th International Symposium*. J. Hollerbach and D. Koditschek (Eds), Springer-Verlag London, (2000) 337–341
9. Full, R.J.: Mechanics and energetics of terrestrial locomotion: From bipeds to polypeds. In: *Energy Transformation in Cells and Animals*. (ed. W. Wieser and E. Gnaiger). Georg Thieme Verlag, Stuttgart. (1989) 175–182 pp
10. Full, R.J. and Koditschek, D.E.: Templates and Anchors — Neuromechanical hypotheses of legged locomotion on land. *J. exp Bio.* **202** (1999) 3325–3332.
11. Full, R.J. and Meijer, K.: Metrics of Natural muscle. In: *Electro Active Polymers (EAP) as Artificial Muscles, Reality Potential and Challenges*. (ed. Y. Bar-Cohen), SPIE & William Andrew/Noyes Publications (2001)
12. Full, R.J. and Tu, M.S.: The mechanics of six-legged runners. *J. exp. Biol.* **148** (1990) 129–146.
13. Full, R.J. and Tu, M.S.: Mechanics of rapid running insects: two-, four-, and six-legged locomotion. J. exp Bio. 156 (1991) 215–231
14. Garland, T., Jr.: Testing the predictions of symmorphosis: conceptual and methodological issues. Pages 40–47 in *Principles of Animal Design: The Optimization and Symmorphosis Debate*, E. R. Weibel, L. Bolis, and C. R. Taylor, eds. Cambridge Univ. Press, Cambridge, U.K. (1998)
15. Greiner, H., Shectman, A., Chikyung Won, Elsley, R. and Beith, P.: Autonomous legged underwater vehicles for near land warfare. *Proceedings of Symposium on Autonomous Underwater Vehicle Technology*, New York, NY: IEEE, (1996). p.41–48
16. Jacob, F.: Evolution and tinkering. *Science* **196** (1977) 1161–1166
17. Kubow, T. M. and R.J. Full. The role of the mechanical system in control: A hypothesis of self-stabilization in hexapedal runners. *Phil. Trans. Roy. Soc. London B.* **354** (1999) 849–862
18. Lipson, H. and Pollack, J.B.: Automatic design and manufacture of robotic life-forms *Nature* **406** (2000) 974–978
19. Martinez, M.M., Full, R.J. and Koehl, M.A.R.: Underwater punting by an intertidal crab: a novel gait revealed by the kinematics of pedestrian locomotion in air versus water. *J. exp Bio.* **201** (1998) 2609–2623
20. Pelrine, R., Kornbluh, R., Qibing Pei and Joseph, J.: High-speed electrically actuated elastomers with strain greater than 100%. *Science* **287** (2000) 836–839
21. Saranli, U., Buehler, M. and Koditschek, D.E.: Design, modeling and Control of a compliant hexapod robot. In *Proc. IEEE Int. Conf. Rob. Aut.* (2000) 2589–2596.
22. Schmitt, J. and Holmes, P.: Mechanical models for insect locomotion: Dynamics and stability in the horizontal plane. I. Theory. In: *Biological Cybernetics* **83** (2000) 501–515
23. Schmitt, J. and Holmes, P.: Mechanical models for insect locomotion: Dynamics and stability in the horizontal plane. II. Application. In: *Biological Cybernetics* **83** (2000) 517–527.
24. Vogel, S.: Cats' Paws and Catapults: Mechanical Worlds of Nature and People. New York: Norton, (1998). 382 pp

Interactions between Art and Mobile Robotic System Engineering

Francesco Mondada and Skye Legon

Autonomous Systems Laboratory (ASL-ISR-DMT)
Signal Processing Laboratory (LTS-DE)
Swiss Federal Institute of Technology (EPFL)
CH-1015 Lausanne
Switzerland
francesco.mondada@epfl.ch
skye.legon@epfl.ch

Abstract. The field of mobile robotics offers a new medium for public entertainment and art. Mobile robots can move, react, and interact in the real world, generating behaviors that can be used as a new artistic medium quite different from sculptures, drawings or video. This new medium, like other technological media such as video or the Internet, requires considerable technical know-how to be exploited successfully. The successful design of a mobile robot demands a strong interdisciplinary and systems-oriented engineering process. The addition of artistic constraints adds a new dimension to the engineering problem and reinforces the need for a coherent approach to the design.
This paper illustrates this interdisciplinary approach with six examples of robotic art and entertainment projects that demonstrate the methodological issues needed for this type of work. Several aspects of the projects are discussed, including the artistic effects on the public, the sometimes problematic interaction between artists and engineers, and details of the mechanical, electronic and behavioral designs as applied to entertainment.

Introduction

Mobile robots are an increasingly popular trend in entertainment and art. International exhibitions such as Hannover 2000 have exploited this, while completely new exhibitions like RoboFesta in Japan are centered entirely on robotics. Large entertainment companies like Sony are investing heavily in robotics, while an increasing number of films and video games feature robots, or are based entirely on robots. Artists as well have discovered robotics as a new medium for their creations, seeing in robotics a means of eliciting strong reactions and emotions in the public. While films and video games admittedly do not need to actually create real robots, most of the other domains involve at least a partial implementation. It is clear that mobile robots are assuming an increasing role in our society in general, in commercial entertainment, and in art as a means of expression. But why?

T. Gomi (Ed.): ER 2001, LNCS 2217, pp. 121–137, 2001.

Mobile robots bring to art and entertainment the ability to elicit very strong projections in the viewer, in the sense that observers project their own interpretation of the robot's behavior onto the robot. This tendency to ascribe intelligence and motivations to the robot that it does not actually possess has been observed by researchers for many years (well known are Penny [1] in art and Braitenberg [2] in psychology), and the authors have confirmed these effects when presenting mobile robots during exhibitions, demonstrations or presentations to the public. The key aspects for generating a successful suspension of disbelief appear to be physical mobility and autonomous behavior. These two aspects are strongly linked in mobile robotics and generate in the viewer the impression of living organisms having their own intelligence. Simple mobility is sufficient to imbue life into the object, while autonomous behavior provides the illusion of intelligence. The behaviors do not need to be very complex, but simply a good mix between unpredictable and understandable.

Other techniques can generate projections in the observer, such as a close mimicry of existing living creatures through form, sound, behavior, and mimicry through anthropomorphism in general [3]. Mobility and autonomous behavior can also be considered in a sense as mimicking living organisms, but the level of mimicry remains very general. Recent efforts have pushed robotic mimicry as far as to simulate domestic animals like cats and dogs. This level of mimicry can yield very strong projections, but can also create expectations, which can have an adverse effect as the observer risks being disappointed when the mimicry does not meet up to expectations.

Most current robotic art and entertainment does not take advantage of the full potential of the field of mobile robotics. Many artists use the popular image of robots as complex technology to hide what are in reality very simple automatons. Others explore the aspect of motion, but are unable to create real behaviors. Very few artists make a real effort to integrate the science of mobile robotics in their projects. In both art and entertainment the robot must be carefully designed in order to achieve a desired projection. The design can be superficially seen as an engineering problem, but the range of competences required to solve the problem goes far beyond classical engineering domains. Mobile robotics is already an interdisciplinary domain where engineers must integrate mechanics, optics, electronics, computer science, and artificial intelligence. The added constraint of evoking specific emotions requires an entirely new set of competences, including design, aesthetics, history, and psychology. By itself, mobile robotics is a field that calls for strong teamwork between engineers from different disciplines; robotics in entertainment and art demands a far greater level of cooperation in a team composed of more disparate backgrounds to efficiently design a coherent end product.

The following article presents several examples of art and entertainment robotics where mobility and behavior are used in such as way as to best exploit the technology itself without pushing real-world mimicry too far. For each project is presented the motivation, the implementation and the design issues involved, as well as lessons learned from what worked and what didn't work quite as well. The projects have all been developed by K-Team, a robotics company based in Morges, Switzerland, manufacturers of the miniature research robot Khepera. All projects but one were commercial ventures, the exception involving an artistic collaboration for an international electronic art exhibition.

1   Example 1: Khepera Advertising Display

The train station in Lausanne provides several displays that companies can rent to advertise their products (see figure 1). The basic problem with these displays is that nobody looks at them, with the exception of the odd traveller passing time while waiting for the next train, and even then they attract but half-hearted interest. Considering the high traffic passing through the station daily, these displays offer a high potential audience, if only a means could be found to attract their attention.

Fig. 1. Advertising displays in the Lausanne train station on the right-hand side of this underground passage.

Previous experience with the Khepera robot had shown that even very simple obstacle avoidance behavior can grab the attention of quite a crowd, and so the idea of an "active" advertising display using a Khepera was developed in conjunction with local designer Krisztina Takacs-Floreano. A company in Lausanne was at this time (1997) marketing Apple's Newton PDA and purchased this idea to animate their display in the train station.

1.1   Concept and Implementation

The display consisted of a Khepera robot moving on a world map featuring a Newton PDA (see figure 2). The world and the mobility of the Khepera were used to symbolize the portability of the PDA. The Khepera robot was powered by a wire connected to a rotating contact on the ceiling, and roamed over the world map in an environment composed of several objects such as a telephone, typewriter and computer, each representing alternatives to the PDA.

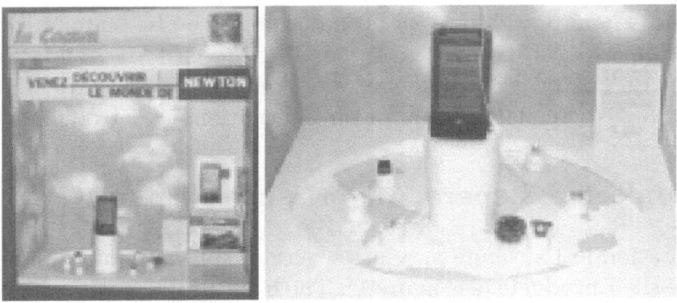

Fig. 2. The "active" display at the Lausanne train station.

Technically speaking, the robot had to simply avoid obstacles in its custom environment for roughly 20 hours per day, seven days per week. The planned duration of the display was for three months but its successful operation was extended to a full six months. The reaction to the display has been very positive and interesting. The animation attracted a large number of people, many of who passed long periods in front of the display. The display was able to catch the attention of both strolling families as well as harried businessmen (see figure 3).

1.2   Lessons Learned

This first project demonstrated clearly how mobility and autonomous behavior can attract an audience. Mobility, especially in a display where people do not expect motion, is a strong attention-getter. Once the observer is attracted and approaches the display, the behavior begins to play a key role: the viewer studies

Fig. 3. A series of visitors demonstrating the efficacy of the display, with one man returning for a closer look after passing by the display.

the robot and tries to understand its behavior. The Khepera displayed very random behavior due to both the noisy environment seen by its optical sensors, and its navigation algorithm designed to keep the robot moving no matter what situation it found itself in. On many occasions observers would start to leave the display, but then turn back to study the robot's next movement, and then another, and another... Children especially could pass long periods fascinated by the robot, effectively trapping their parents in front of the exhibit until they were reluctantly pulled away.

From a design point of view, it was K-Team's first development that was dictated by non-technical constraints. The collaboration with the designer was very successful and the communication between both parties passed smoothly. The technical aspect, however, proved far more difficult. The behavior and environment, though apparently simple, were far more complex than imagined due to the power supply cable and the central support for the Newton. A great deal of trial-and-error was required before finding a suitable position and shape for the Newton's support such that the cable would not catch on the support and trap the robot.

2   Example 2: Robot Avatar Dreaming with Virtual Illusions

This second project was developed by the artist Franz Fischnaller and his group FABRICATORS in Italy, and featured the use of a Koala robot from K-Team. Fischnaller has extensive experience in exploiting new technology for artistic projects, especially works involving virtual reality [4]. He contacted K-Team for support in integrating mobile robotics into an exhibition, and K-Team joined the effort as a robotics consultant. Fischnaller merged his artistic ideas with K-Team's robotics technology for his work Robot Avatar Dreaming with Virtual Illusions, certainly the most abstract and artistic of the projects with which K-Team has been involved. It is also K-Team's only non-commercial venture of the six examples presented in this paper, though compensation came in the form of an honorary mention at Ars Electronica 99, a major international event in electronics art.

2.1   Concept and Implementation

This project combined virtual reality with real robotics, including local interaction using a joystick and remote interaction via the Internet. The robot's real-world environment was a sand arena illuminated by lateral light to emphasize the irregularity of the sand and create a suggestive and enigmatic environment, as well as to show the path of the robot (see figure 4 left). The Koala robot was modified with a transparent circular plate to generate interesting light effects, described by the authors as "a smart skin which is at the same time a catalyzer and mirror of his experience and emotions which he lives by means of the interactivity" (see figure 4 right).

Fig. 4. The Koala in its arena.

The robot had its own autonomous behaviors which were associated with the actions of the avatar in the virtual world and modified via interaction with the visitors. One interesting aspect of the association between robot and avatar is that the robot could avoid obstacles seen only by the avatar in one of several virtual worlds (shown in figure 5), resulting in enigmatic behavior to visitors only observing the real world. For more information please visit [5].

Fig. 5. Two of the worlds explored by the robot avatar.

2.2   Lessons Learned

Mobility and autonomous behavior both played key roles in this work, and were also symbolic: the physical robot and its mobility are symbols for the real, visible world, while the behavior serves as a link between the physical and virtual worlds, as well as between the robot/avatar and visitor.

In the context of this paper, the most interesting aspect of this work was the nature of the interaction with Fischnaller and his group. K-Team did not perform any development for the project and remained simply a consulting partner

(K-Team was more deeply involved in Fischnaller's Pinocchio project, presented as example 4). The robotics part of the project was actually a fairly small part of the work as a whole, but it is nonetheless significant that Fischnaller and his group programmed the robot themselves without any prior robotics experience! Although the participation of K-Team was limited to consulting, the quality of the work and the successful interaction make it a model for interdisciplinary projects. K-Team made a great effort to understand the goals and artistic constraints of the project, and propose suitable techniques to attain the desired effect. The artistic team too worked hard to understand the limits of the technology and adapt their artistic vision appropriately. Thus through a relationship of mutual respect for each other's fields was it possible to find solutions together that were both technically feasible and artistically coherent. In contrast, most partners seeking robotics technology underestimate the technical obstacles involved, and arrive with high expectations and precise demands, without any margin for adapting their concept to the limits and capabilities of mobile robots. Fischnaller fully integrated the suggestions of K-Team into his project, even the artistic ones! The end result was very successful, earning an honorary mention at the prestigious electronic art exhibition Ars Electronica 99, demonstrating the efficacy of this design methodology.

3   Example 3: Ball Sorting Demonstration

This purely commercial project was developed for a design firm charged with creating a robotics exhibit for an expo organized by their client, a large multinational automotive firm. The designers approached K-Team for support in creating an attractive demonstration, and together a concept was developed whereby three Khepera robots would cooperate to "harvest" colored balls in a small arena.

3.1   Concept and Implementation

A circular arena roughly two meters in diameter was divided with painted lines into three equal wedges, each the territory of one of the three robots (see figure 6). Balls of three different sizes corresponding to each robot were introduced into the arena from the center and would end up randomly in one of the three areas. The robots roamed their terrain searching for balls, and when found would grab the ball with a gripper device. Noting the size of the ball, the robot would decide if it belonged to him, and if so then hunt for his "nest" and deposit the ball in a hole. If the ball belonged to one of his friends, he would roll over to the appropriate border between their two territories and fling the ball towards his friend, who would find the ball on his own in due course. In this roundabout way the three robots could quite quickly clear the arena of a large number of balls (see figure 7).

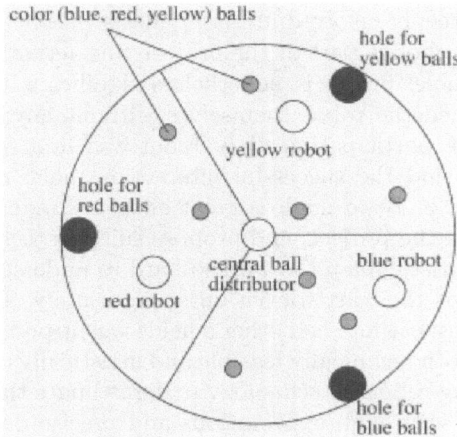

Fig. 6. The layout of the arena.

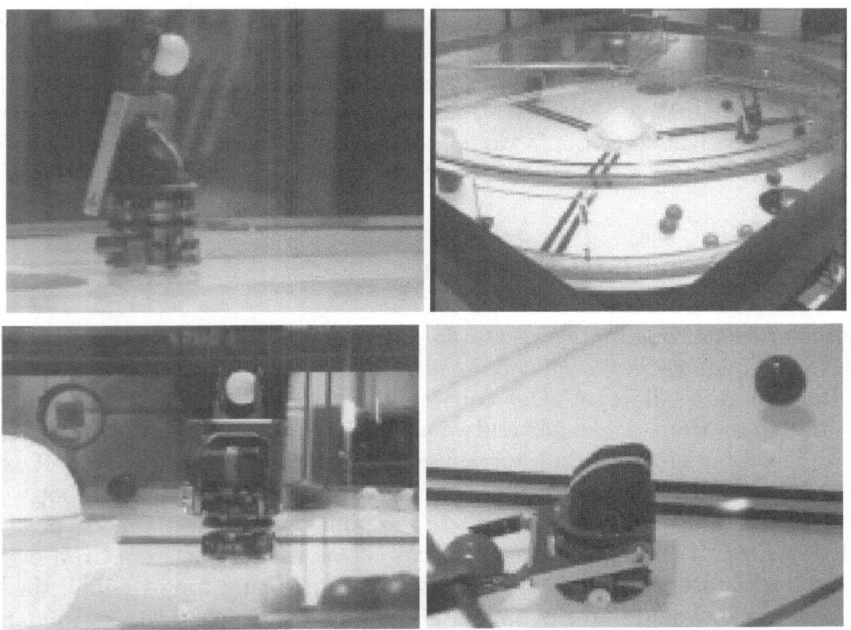

Fig. 7. Robots working together to harvest colored balls.

3.2   Lessons Learned

Here too mobility and autonomous behavior played key roles: the mobility attracted visitors to the exhibit, whereupon they remained rooted to the spot, fascinated by the robots' behavior and trying to deduce their modus operandi. The goal of the designers was thus achieved with success, and the exhibit became one of the main attractions of the entire exhibition.

This success was achieved, however, despite an often-difficult interaction with the design team. From the start of the project they rigidly defined the purpose of the installation and a list of inviolable constraints. By means of example, one such constraint was that the robots must be completely wireless, as a cable was seen to undermine the concept of computational and behavioral autonomy. This caused large difficulties, as the robots were required to run continuously 8–12 hours a day, and battery power supply remains a thorny issue in mobile robotics. After significant development the project was in its final tests just before the start of the exhibition, and to aid with their debugging the robots were connected to a computer via a cable so that the robots could display their "thought processes" on the screen. An important representative from the automotive firm happened to pass by and was fascinated by the information displayed on the screen, as it allowed him to see how the robot worked. He asked why this information wasn't part of the exhibit, and was politely told that it was difficult to transmit this information without a cable. His response was maddening: "Why not use a cable, then?" This anecdote demonstrates how a top-down design approach with no consideration for low-level technical constraints can result in absurd situations and inefficient designs.

4   Example 4: Pinocchio Interactive

Pinocchio Interactive is another project created by Franz Fischnaller and FAB-RICATORS integrating robotics, virtual reality and user interaction. It allows the visitor to explore a modern version of the classical story of Pinocchio by using a joystick to control a real robotic Pinocchio puppet or its virtual counterpart.

4.1   Concept and Implementation

The 1.8m tall Pinocchio puppet is suspended, like a real puppet, by wires that control its movements. It hangs before a 3x3m screen showing its virtual alter ego and the virtual story (see figures 8–10). The real and virtual Pinocchio puppets can interact with each other, and can be manipulated as well by the visitors using a joystick. More information can be found at [6].

The first Pinocchio puppet (still in the prototype phase) is controlled by five motors, each motor controlling one or more cables connected to its arms and legs, and is thus animated in the same manner as a classical puppet.

Fig. 8. The real Pinocchio puppet and its alter ego in the virtual world.

Fig. 9. The Pinocchio puppet placed in front of the screen showing the virtual world.

Fig. 10. Images from the virtual Pinocchio story.

4.2   Lessons Learned

Mobility and behavior are again present in this work, drawing the visitor into the story using a variety of actions and animations of the puppet interacting with its virtual alter ego.

Aside from difficulties in finding financing for this type of project, this experience again was a highly positive one for the same reasons as example 2, and serves as a demonstration for optimal interaction between engineers and artists.

5   Example 5: Flower Pot

This project is a permanent animated sculpture located on the roof of a building at a technical school in Neuchâtel in Switzerland. As part of a renovation project the school consecrated a part of their budget to "beautify" an unsightly low gravel-roofed building in the middle of their campus. The winners of the design competition were a group of local artists, the "gruppo GPM" of Pierre Gattoni, Emmanuel Du Pasquier and Yvo Mariotti, who made contact with K-Team for aid with the implementation of the robotics of their animated sculpture.

5.1   Concept and Implementation

The animated sculpture is in reality an enormous autonomous flowerpot with the same proportions as a normal flowerpot, but with a diameter of 3 meters and weighing several hundred kilograms. In place of flowers the pot supports a 25 meter high flexible mast, designed to sway artistically in the breeze (see figure 11). The upper surface of the pot is covered with solar panels to collect energy for the internal batteries, which in turn power two large drive motors while six castor wheels ensure the stability of the system. The base of the pot is ringed with several infrared distance sensors that enable the pot to avoid obstacles, and inside the pot is hidden a reservoir and a pump to collect rainwater and to occasionally shoot plumes of water into the air.

The flowerpot navigates around the roof avoiding obstacles (skylights and the external walls) several hours per day. An important characteristic of the system is its rather modest speed of 2 to 5 meters per hour. This creates an interesting effect: when looking at the pot it seems immobile, but each time you look it is in a different location (see figure 12). The artists draw an analogy with the moon: you do not see it move, but you always have to look around to find it again.

From a technical standpoint, the robot behaves like a (very) large Khepera robot, and is controlled by K-Team's Kameleon board with a custom extension to read the infrared sensors and control the motors. Developing a robust classical navigation algorithm with such simple sensors proved impossible, and so a Braitenberg-inspired neural network was used, with sensor parameters and positioning evolved using a genetic algorithm. The custom genetic algorithm was developed using the simulation software SysQuake (see figure 13).

Fig. 11. The flowerpot on the roof amid the snows of winter.

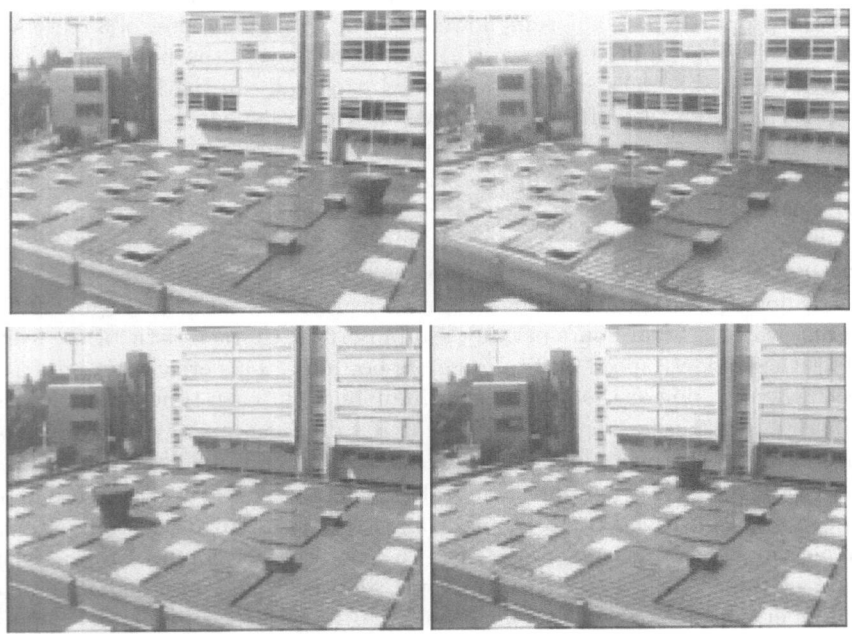

Fig. 12. The flowerpot moving around the roof over a period of 24 hours.

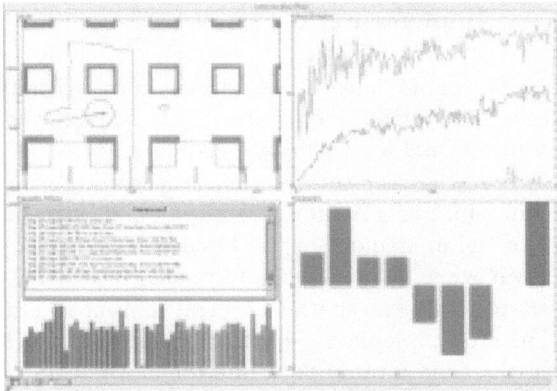

Fig. 13. Calculation of robot parameters using genetic algorithms and the mathematical simulation software SysQuake. The top-left of the image shows a map of part of the roof and the simulated trajectory of the robot.

5.2   Lessons Learned

As with the other projects, mobility and autonomous behavior are key elements, but differ from the others, however, in that these two attributes are partially hidden, or made enigmatic, by the temporal scaling of the behavior.

As K-Team was contacted after the project was defined, the interaction with the artists was limited to modification of minor details, prohibiting proper bottom-up development. In addition, as the project progressed, structural considerations due to the size of the mast required an increasingly larger and heavier pot, greatly reducing the free space between the pot and obstacles, and thus rendering the navigation problem exceedingly difficult. Interdisciplinary interaction for the development of the installation was clearly insufficient, yielding a seemingly never-ending series of technical hurdles.

6   Example 6: Robot Theater

This project is the most complex development in entertainment robotics undertaken by K-Team. It is a robot theatre, consisting of four robots playing the roles of mother, father, son and daughter. The project was developed for a Swiss exhibition called "Little Children: Joy and Exasperation" organized by Ethno-Expo to run from the year 2000 through 2003. The family of robots acts out a series of short scenes of domestic life, some finishing happily, and others not so happily, with the goal of educating young children and their families on the joys and frustrations of family life.

6.1   Concept and Implementation

The custom-designed robots have a cylindrical body with two hidden drive wheels, simple arms, a motorized mouth, two LED-matrix eyes and one "heart" on the body, and an internal loudspeaker (see figure 14).

The robots act out the scenes on a stage representing an apartment with kitchen, living room, children's bedroom, and parents' bedroom (see figure 15). The "front door" of the apartment in the kitchen leads to a small closet space where the robots rest when not onstage and recharge their batteries.

The entire system is fully automatic: a central computer controls the stage accessories (television, stove, telephone, door, rotating table, the sound and lights) and controls each robot via a radio data link for movements and a radio sound channel for the voice. Each robot is constructed around a Kameleon controller board, which communicates with the host computer via a radio modem, and controls four motors (two wheels and two arms), the mouth servomotor, the LED-matrix displays, and several contact-switch sensors (see figure 16).

Challenging aspects of the project included the positioning of the robots and the scripting of the scenes. Robot positioning was performed using internal odometry with regular recalibrations using mechanical landmarks, which were carefully integrated into the script and did not disturb the spectators' appreciation of the presentation. The scripting of the scenes was a complex task, and a custom multi-tasking event-based scripting language was developed for this purpose. Scripts were sent over the radio link to the robots and compiled on the fly to permit the coordination of the robot's complex actions, involving physical displacement, moving its arms and mouth, and controlling the LED-matrix

Fig. 14. Robinette and her father, two of the four robots in the family.

Fig. 15. Some scenes from the robot performance.

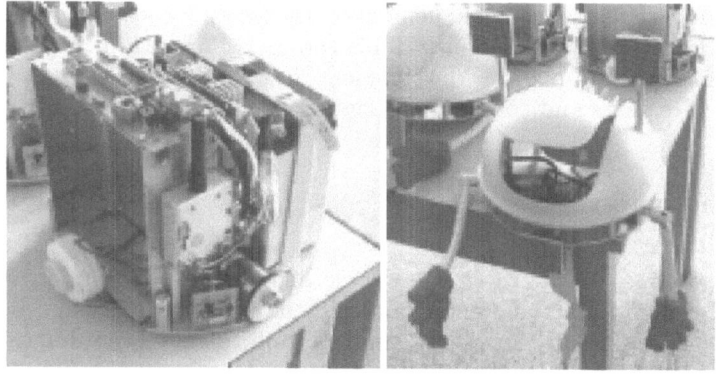

Fig. 16. Robot chassis with body removed and the head and arms of the robot.

displays. Equally importantly, the system permitted the coordination of actions between several robots, resulting in fully synchronous collective behaviors.

6.2   Lessons Learned

Unlike the other projects, the robots in this project were not behaviorally autonomous, but this was compensated for by the complexity of the scripted actions. The mobility and expressions of the robots are by far the richest of the examples presented here, and evoke strong projections of family life in both parents and children. However, as explained in the introduction, this mimicry is a double-edged sword, and the successful representation of family life also creates expectations on the part of the viewers. For example, the relatively slow speed of the robots can disappoint visitors, who project human speeds onto the robots, despite the obvious physical dissimilarity.

The design process involved a rich interaction between the artistic and technical teams, necessitated by the size and ambitious nature of the project. The interaction was facilitated by Swiss writer and researcher Jean-Bernard Billeter, whose background in both theatre and robotics aided in bridging the gap between the two teams. Aside from writing the scripts for the project, his efforts to explain technical constraints to the artistic team (and to sensitize a team of engineers to subtle artistic considerations) did much to smooth development.

This ambitious project demanded from the outset special efforts to link the artistic conception with the engineering design. The degree of cooperation required was at first greatly underestimated and only became clear once significant problems arose. A notable example is that of the scripting language: K-Team had planned to provide a simple linear time-based scripting language, but as development progressed it became evident that it would be wholly insufficient for coordinating a complex series of simultaneous actions of variable duration both on the robot itself, and for the scene as a whole. It was soon realized that anything less than a completely general multi-agent, event-based, multi-tasking scripting language would seriously limit the artistic possibilities. This necessitated an unforeseen and costly development that underlined the need to properly define requirements and coordinate design efforts from the bottom-up.

7   Conclusion

The six examples presented in this article demonstrate two important aspects of robotic design for art and entertainment: the use of mobility and behavior to generate projections in the viewer, and the difficulties of coordinating artistic and technical teams to achieve this goal.

The two key elements in successfully generating emotional projections in the observer are motion and autonomous behavior. Motion is important for imbuing the robot with the illusion of life, while autonomous behavior allows the viewer to project needs, wants, and motivations onto the robot, thus bestowing it with a perceived intelligence. These projections can be used to simply entertain the visitor (as in examples 1 and 4) or if placed in a specific context (as in examples 2, 3, 5 and 6) can be employed to convey a message.

The nature of the interaction between the artistic/design team and the robotics/engineering team is critical for the success of this type of development. Careful thought must be given to the design process, in particular the coordination of a joint bottom-up development between the two teams. Classical top-down client-provider relationships between artists and engineers can result in an inefficient design process and an unsuccessful end product. Engineers must be prepared for the unrealistic expectations of non-specialists, while artists must accept that mobile robotics remains a specialized field that cannot be easily adapted to suit every need. The success of the project demands that both parties strive to understand the other's constraints and are prepared to collaborate in developing a mutually satisfactory design.

References

1. S. Penny, "Embodied Cultural Agents: at the intersection of Art, Robotics and Cognitive Science", *AAAI Socially Intelligent Agents Symposium*, MIT, November 8–10, 1997
2. V. Braitenberg, *Vehicles: Experiments in Synthetic Psychology*. Cambridge, MIT Press, 1984
3. S. Penny, "Why do we want our machines to seem alive?", *Scientific American, 150th anniversary issue*, September 1995
4. http://www.fabricat.com
5. http://www.fabricat.com/ROBO_HTM/robot.html
6. http://www.fabricat.com/pinoc_home.html

Author Index

Lecture Notes in Computer Science

For information about Vols. 1–2126
please contact your bookseller or Springer-Verlag

Vol. 2168: L. De Raedt, A. Siebes (Eds.), Principles of Data Mining and Knowledge Discovery. Proceedings, 2001. XVII, 510 pages. 2001. (Subseries LNAI).

Vol. 2170: S. Palazzo (Ed.), Evolutionary Trends of the Internet. Proceedings, 2001. XIII, 722 pages. 2001.

Vol. 2172: C. Batini, F. Giunchiglia, P. Giorgini, M. Mecella (Eds.), Cooperative Information Systems. Proceedings, 2001. XI, 450 pages. 2001.

Vol. 2173: T. Eiter, W. Faber, M. Truszczynski (Eds.), Logic Programming and Nonmonotonic Reasoning. Proceedings, 2001. XI, 444 pages. 2001. (Subseries LNAI).

Vol. 2174: F. Baader, G. Brewka, T. Eiter (Eds.), KI 2001: Advances in Artificial Intelligence. Proceedings, 2001. XIII, 471 pages. 2001. (Subseries LNAI).

Vol. 2175: F. Esposito (Ed.), AI*IA 2001: Advances in Artificial Intelligence. Proceedings, 2001. XII, 396 pages. 2001. (Subseries LNAI).

Vol. 2176: K.-D. Althoff, R.L. Feldmann, W. Müller (Eds.), Advances in Learning Software Organizations. Proceedings, 2001. XI, 241 pages. 2001.

Vol. 2177: G. Butler, S. Jarzabek (Eds.), Generative and Component-Based Software Engineering. Proceedings, 2001. X, 203 pages. 2001.

Vol. 2180: J. Welch (Ed.), Distributed Computing. Proceedings, 2001. X, 343 pages. 2001.

Vol. 2181: C. Y. Westort (Ed.), Digital Earth Moving. Proceedings, 2001. XII, 117 pages. 2001.

Vol. 2182: M. Klusch, F. Zambonelli (Eds.), Cooperative Information Agents V. Proceedings, 2001. XII, 288 pages. 2001. (Subseries LNAI).

Vol. 2183: R. Kahle, P. Schroeder-Heister, R. Stärk (Eds.), Proof Theory in Computer Science. Proceedings, 2001. IX, 239 pages. 2001.

Vol. 2184: M. Tucci (Ed.), Multimedia Databases and Image Communication. Proceedings, 2001. X, 225 pages. 2001.

Vol. 2185: M. Gogolla, C. Kobryn (Eds.), «UML» 2001 – The Unified Modeling Language. Proceedings, 2001. XIV, 510 pages. 2001.

Vol. 2186: J. Bosch (Ed.), Generative and Component-Based Software Engineering. Proceedings, 2001. VIII, 177 pages. 2001.

Vol. 2187: U. Voges (Ed.), Computer Safety, Reliability and Security. Proceedings, 2001. XVI, 261 pages. 2001.

Vol. 2188: F. Bomarius, S. Komi-Sirviö (Eds.), Product Focused Software Process Improvement. Proceedings, 2001. XI, 382 pages. 2001.

Vol. 2189: F. Hoffmann, D.J. Hand, N. Adams, D. Fisher, G. Guimaraes (Eds.), Advances in Intelligent Data Analysis. Proceedings, 2001. XII, 384 pages. 2001.

Vol. 2190: A. de Antonio, R. Aylett, D. Ballin (Eds.), Intelligent Virtual Agents. Proceedings, 2001. VIII, 245 pages. 2001. (Subseries LNAI).

Vol. 2191: B. Radig, S. Florczyk (Eds.), Pattern Recognition. Proceedings, 2001. XVI, 452 pages. 2001.

Vol. 2192: A. Yonezawa, S. Matsuoka (Eds.), Metalevel Architectures and Separation of Crosscutting Concerns. Proceedings, 2001. XI, 283 pages. 2001.

Vol. 2193: F. Casati, D. Georgakopoulos, M.-C. Shan (Eds.), Technologies for E-Services. Proceedings, 2001. X, 213 pages. 2001.

Vol. 2194: A.K. Datta, T. Herman (Eds.), Self-Stabilizing Systems. Proceedings, 2001. VII, 229 pages. 2001.

Vol. 2195: H.-Y. Shum, M. Liao, S.-F. Chang (Eds.), Advances in Multimedia Information Processing – PCM 2001. Proceedings, 2001. XX, 1149 pages. 2001.

Vol. 2196: W. Taha (Ed.), Semantics, Applications, and Implementation of Program Generation. Proceedings, 2001. X, 219 pages. 2001.

Vol. 2197: O. Balet, G. Subsol, P. Torguet (Eds.), Virtual Storytelling. Proceedings, 2001. XI, 213 pages. 2001.

Vol. 2198: N. Zhong, Y. Yao, J. Liu, S. Ohsuga (Eds.), Web Intelligence: Research and Development. Proceedings, 2001. XVI, 615 pages. 2001. (Subseries LNAI).

Vol. 2199: J. Crespo, V. Maojo, F. Martin (Eds.), Medical Data Analysis. Proceedings, 2001. X, 311 pages. 2001.

Vol. 2200: G.I. Davida, Y. Frankel (Eds.), Information Security. Proceedings, 2001. XIII, 554 pages. 2001.

Vol. 2201: G.D. Abowd, B. Brumitt, S. Shafer (Eds.), Ubicomp 2001: Ubiquitous Computing. Proceedings, 2001. XIII, 372 pages. 2001.

Vol. 2202: A. Restivo, S. Ronchi Della Rocca, L. Roversi (Eds.), Theoretical Computer Science. Proceedings, 2001. XI, 440 pages. 2001.

Vol. 2204: A. Brandstädt, V.B. Le (Eds.), Graph-Theoretic Concepts in Computer Science. Proceedings, 2001. X, 329 pages. 2001.

Vol. 2205: D.R. Montello (Ed.), Spatial Information Theory. Proceedings, 2001. XIV, 503 pages. 2001.

Vol. 2206: B. Reusch (Ed.), Computational Intelligence. Proceedings, 2001. XVII, 1003 pages. 2001.

Vol. 2207: I.W. Marshall, S. Nettles, N. Wakamiya (Eds.), Active Networks. Proceedings, 2001. IX, 165 pages. 2001.

Vol. 2208: W.J. Niessen, M.A. Viergever (Eds.), Medical Image Computing and Computer-Assisted Intervention – MICCAI 2001. Proceedings, 2001. XXXV, 1446 pages. 2001.

Vol. 2209: W. Jonker (Ed.), Databases in Telecommunications II. Proceedings, 2001. VII, 179 pages. 2001.

Vol. 2210: Y. Liu, K. Tanaka, M. Iwata, T. Higuchi, M. Yasunaga (Eds.), Evolvable Systems: From Biology to Hardware. Proceedings, 2001. XI, 341 pages. 2001.

Vol. 2211: T.A. Henzinger, C.M. Kirsch (Eds.), Embedded Software. Proceedings, 2001. IX, 504 pages. 2001.

Vol. 2212: W. Lee, L. Mé, A. Wespi (Eds.), Recent Advances in Intrusion Detection. Proceedings, 2001. X, 205 pages. 2001.

Vol. 2213: M.J. van Sinderen, L.J.M. Nieuwenhuis (Eds.), Protocols for Multimedia Systems. Proceedings, 2001. XII, 239 pages. 2001.

Vol. 2215: N. Kobayashi, B.C. Pierce (Eds.), Theoretical Aspects of Computer Software. Proceedings, 2001. XV, 561 pages. 2001.

Vol. 2217: T. Gomi (Ed.), Evolutionary Robotics. Proceedings, 2001. XI, 139 pages. 2001.